GIS FOR HOUSING AND URBAN DEVELOPMENT

Committee on Review of Geographic Information Systems Research and
Applications at HUD: Current Programs and Future Prospects
Committee on Geography
Board on Earth Sciences and Resources
Division on Earth and Life Sciences

NATIONAL RESEARCH COUNCIL
OF THE NATIONAL ACADEMIES

THE NATIONAL ACADEMIES PRESS
Washington, D.C.
www.nap.edu

THE NATIONAL ACADEMIES PRESS, 500 Fifth Street, N.W., Washington, DC 20001

NOTICE: The project that is the subject of this report was approved by the Governing Board of the National Research Council, whose members are drawn from the councils of the National Academy of Sciences, the National Academy of Engineering, and the Institute of Medicine. The members of the committee responsible for the report were chosen for their special competences and with regard for appropriate balance.

This study was supported by Contract No. O-OPC-21952 Task Order One between the National Academy of Sciences and the U.S. Department of Housing and Urban Development. Any opinions, findings, conclusions, or recommendations expressed in this publication are those of the author(s) and do not necessarily reflect the views of the organizations or agencies that provided support for the project.

International Standard Book Number 0-309-08874-7

Additional copies of this report are available from

the National Academies Press
500 Fifth Street, N.W.
Box 285
Washington, DC 20055
(800) 624-6242
(202) 334-3313 (in the Washington metropolitan area)
Internet, http://www.nap.edu

Cover: *Background:* Share of home mortgage applications in Atlanta, Georgia without race-ethnicity information, 1999. Pattern confirms that nondisclosure rates are highest in predominantly African-American neighborhoods. SOURCE: Wyly, E. K., and S. R. Holloway. 2002. The Disappearance of Race in Mortgage Lending. Economic Geography 78(2):129-163
Top left: Urban rowhouse in Washington, D.C. SOURCE: Monica Lipscomb, Washington, D.C.
Center left: GIS made available via community data centers can engage citizens in urban and regional planning as demonstrated by this photograph. SOURCE: Photodisc stock photos, Technology at Work. Copyright 1999 CORBIS.
Bottom right: Examples of data layer overlays including (from top to bottom) land ownership, demographics, transportation and aerial imagery. SOURCE: National States Geographic Information Council and Federal Geographic Data Committee. nd. Using Geography to Advance the Business of Government: The Power of Place to Support Decision Making. CD-ROM. Washington, D.C.: NSGIC.

THE NATIONAL ACADEMIES
Advisers to the Nation on Science, Engineering, and Medicine

The **National Academy of Sciences** is a private, nonprofit, self-perpetuating society of distinguished scholars engaged in scientific and engineering research, dedicated to the furtherance of science and technology and to their use for the general welfare. Upon the authority of the charter granted to it by the Congress in 1863, the Academy has a mandate that requires it to advise the federal government on scientific and technical matters. Dr. Bruce M. Alberts is president of the National Academy of Sciences.

The **National Academy of Engineering** was established in 1964, under the charter of the National Academy of Sciences, as a parallel organization of outstanding engineers. It is autonomous in its administration and in the selection of its members, sharing with the National Academy of Sciences the responsibility for advising the federal government. The National Academy of Engineering also sponsors engineering programs aimed at meeting national needs, encourages education and research, and recognizes the superior achievements of engineers. Dr. Wm. A. Wulf is president of the National Academy of Engineering.

The **Institute of Medicine** was established in 1970 by the National Academy of Sciences to secure the services of eminent members of appropriate professions in the examination of policy matters pertaining to the health of the public. The Institute acts under the responsibility given to the National Academy of Sciences by its congressional charter to be an adviser to the federal government and, upon its own initiative, to identify issues of medical care, research, and education. Dr. Harvey V. Fineberg is president of the Institute of Medicine.

The **National Research Council** was organized by the National Academy of Sciences in 1916 to associate the broad community of science and technology with the Academy's purposes of furthering knowledge and advising the federal government. Functioning in accordance with general policies determined by the Academy, the Council has become the principal operating agency of both the National Academy of Sciences and the National Academy of Engineering in providing services to the government, the public, and the scientific and engineering communities. The Council is administered jointly by both Academies and the Institute of Medicine. Dr. Bruce M. Alberts and Dr. Wm. A. Wulf are chairman and vice chairman, respectively, of the National Research Council.

www.national-academies.org

Acknowledgments

This report has been reviewed by individuals chosen for their diverse perspectives and technical expertise, in accordance with procedures approved by the NRC's Report Review Committee. The purpose of this independent review is to provide candid and critical comments that will assist the authors and the NRC in making their published report as sound as possible and to ensure that the report meets institutional standards for objectivity, evidence, and responsiveness to the study charge. The content of the review comments and draft manuscript remain confidential to protect the integrity of the deliberative process. We wish to thank the following individuals for their participation in the review of this report:

Libby Clapp, Office of the Chief Technology Officer, Government of the District of Columbia, Washington, D.C.
Joseph Ferreira, Massachusetts Institute of Technology, Cambridge
Edward G. Goetz, University of Minnesota, Minneapolis
Sharon Krefetz, Clark University, Worchester, Massachusetts
Harold Wolman, George Washington University, Washington, D.C.

Although the individuals listed above have provided many constructive comments and suggestions, they were not asked to endorse the conclusions or recommendations, nor did they see the final report before its release. The review of this report was overseen by John S. Adams, Department of Geography, University of Minnesota, Minneapolis. Appointed by the National

Research Council, he was responsible for making certain that an independent examination of this report was carried out in accordance with institutional procedures and that all review comments were carefully considered. Responsibility for the final content of this report rests entirely with the authoring committee and the NRC.

Preface

Rapid development of technology has become matter of fact in our daily lives. With the increasing speed of computers, reduction in their cost, and regular development of new software, computers are becoming more accessible and useful to everyday Americans. We begin to see computers as a regular and unremarkable part of our daily lives, however; the pace of integration of new technologies into our organizational structures, public and private, is much slower than the development of those technologies. Nowhere is this clearer than in the development and utilization of geographic information systems (GIS). The potential of GIS to inform housing and urban research and applications is the subject of this report.

GIS is software that uses geographic (spatial) location as the organizing principle for collection, storage, analysis, and presentation of information in digital form. It began as a tool for planning, moved forward into engineering through CAD (computer aided drafting), and has rapidly developed into the best enterprise software available for management and decision support. In the past 20 years, GIS has developed rapidly, increasing its potential for effective use in both public and private organizations. However, development of effective enterprise uses of GIS and creation of a national infrastructure supporting its use have been slow.

With the work of the Federal Geographic Data Committee (FGDC) and all the federal, state, and local participants in their work, the concept of the national spatial data infrastructure (NSDI) has begun to move from an idea to reality. The U.S. Department of Housing and Urban Development (HUD) has a role to play in the FGDC, specifically in the development of data about urban areas. HUD's mission is to promote adequate and affordable housing, economic opportunity, and a suitable living environment free from discrimination for all Americans (HUD, 2002). The daily work of the department fosters the use of GIS and the development of locational data in the housing agencies throughout the country. HUD's Office of Policy Development and Research is central to this effort.

HUD asked the Committee on Geography of the National Research Council to review its work and that of its local housing agencies in the development and use of GIS. The agency asked for recommendations to maximize the quality and use of its information, and that of the local housing agencies with which it works.

To address this charge, the Committee to Review Research and Applications of GIS at HUD: Current and Future Programs held three meetings between February 2002 and July 2002. These meetings included testimony from HUD staff and other experts in GIS applications in areas of neighborhood change and discrimination, housing, and metropolitan research and technical tools needed for effective dissemination and data accuracy. As background, the committee reviewed relevant HUD documents, pertinent National Research Council reports, and other literature and technical reports, and engaged in discussion with other federal agencies whose responsibilities include urban and community issues.

This report is written for multiple audiences. The network of people who use HUD's data for policy and research purposes is a broad community: professors and students at colleges and universities, policy makers and analysts working for local governments, HUD program managers and research scientists, and neighborhood leaders and residents employed by community-based organizations. The committee heard presentations from and interviewed representatives from the U.S. Bureau of the Census, the U.S. Department of Transportation, the Federal Geographic Data Committee, and public, non-profit, and private sector groups at the national, state, and local levels.

I would like to thank all the members of the ad hoc committee, the workshop contributors and presenters, and the National Research Council staff for their efforts in creating this report. In addition, I thank the Committee on Geography for the opportunity to serve and be involved in this effort.

Eric A. Anderson
Chair

Contents

Executive Summary

It is 2007 and a large public housing complex in Our Town is slated for demolition. Local officials consider alternative response initiatives for their city, including a community center for information, computer and job skills training, and a safe place for residents to bring their children to play and learn. Prospective residents will need information about the availability of new housing and transportation to jobs and childcare, health care, and other family services. They can get this information from the local library that provides access to HUD USER[1] and to a related newsletter with up-to-date information on urban and community issues, as well as from other federal, state, and local government agencies, and private Internet data sources. City officials use these resources to gather the information they need to decide the best course. The information they gather includes lessons learned from previous housing initiatives about neighborhood impacts, the influence of regional housing markets on Section 8 voucher programs, and the implications of poverty concentration. They use tools available online to create maps to display this information, and they bring these maps to a town meeting about planning the new housing and community facilities.

These data and tools are available in part because the U.S. Department of Housing and Urban Development's (HUD) Office of Policy Development

---

[1]An online source of HUD data and related information. See <http://www.huduser.org>.

and Research (PD&R) uses geographic information systems (GIS)[2] to analyze neighborhood change and market trends affecting the housing market, and to understand the relationship between transit and housing in lifting people out of poverty. Research done by HUD program offices, such as PD&R, has determined factors that contribute to the overall effectiveness of various housing programs, and characteristics of neighborhoods requiring incentives to promote the use of Section 8 low-income vouchers. HUD research is aimed at developing planning measures to meet the varied needs of dislocated residents and to improve the dispersion of low-income residents throughout the city and region.

HUD and the local public housing officials work with partners in the Department of Health and Human Services and the Bureau of Labor Statistics to plan the best course of action. Local officials use GIS to determine where community block grants and new community facilities are most needed. The public housing residents had already used an online GIS program at their local housing authority. They understood that GIS can be used to consider housing options under the Section 8 program based on criteria that are important to Our Town residents, such as the vacancies' proximity to quality schools, public transportation, entry-level employment opportunities, and special medical or social services. As a result of the information they have obtained from HUD, the local housing officials feel confident that investment in their chosen initiative will help provide decent, safe, and affordable housing, and support a safe and prosperous community.

HUD AND A CHANGING URBAN AMERICA

The story of Our Town highlights HUD's role in providing data and information for communities across the nation. Created in 1965 to address civil rights, urban poverty, and the state of American cities, HUD is an agency with a mission to increase homeownership, support community development, and increase access to affordable housing free from discrimination (HUD, 2002). Many of the issues and problems that HUD must address are geographic in nature pertaining to: location, e.g., of housing and jobs; spatial relationships, e.g., among a neighborhood, a city, and a region; and the qualities of place, e.g., patterns of crime or environmental quality. To carry out their mission and to address complex issues of urban poverty and the declining state of American cities, HUD needs to collect and disseminate relevant data, carry out research, and partner effectively with other actors in the urban arena.

---

[2]GIS is computer-based system for the collection, storage, analysis, and output of information that is spatially referenced (Obermeyer and Pinto, 1994).

This will require technical capacity in the application of geographic information and considerable research expertise in spatial analysis.

Although poverty occurs in higher proportions in rural areas than in urban areas, a much greater proportion of the U.S. population lives in cities. Therefore, poverty in the United States is largely urban, the product of processes working at the metropolitan and regional level. Economic growth at the regional level can raise housing costs in inner cities, shift investments from the city to the suburbs, and draw away the jobs that once provided income to the urban poor. Poverty in the central city has strong spatial characteristics. The poor are often spatially segregated from the middle class and physically removed from basic services, such as health care, childcare, and retail sales, and from cultural amenities such as libraries and museums. Both the percentage of inner city neighborhoods that are poor and the percentage of poor people living in those neighborhoods have risen in recent decades. Similarly, although poverty rates have declined for certain groups, the overall income gap between the rich and the poor is widening.

The nature of urban and regional dynamics and of land and housing markets demand the deployment of modern data management techniques and analysis tools. Geographic tools and spatial analysis are useful to HUD for assessment of program effectiveness, understanding housing needs, and addressing broader issues of residential segregation and poverty, as well as infrastructure provision and access to services. GIS is a computer tool for understanding where things are located on the surface of the Earth. More than simply drawing maps, GIS allows analysts and citizens to answer questions such as: "Where is assisted housing located in my neighborhood?" "Which environmental hazards are within 5 miles of where I live?" "How can I get from where I live to where I can work?" "What is the best location for a day-care facility in my community?" GIS does this by permitting the integration and analysis of different kinds of data (e.g., environmental data and demographic data) and the display of the results in a visually attractive and graphically explanatory way. Used as a tool for data management and spatial analysis, the information derived from GIS is as accurate as the data that went into the system and as relevant as the questions posed.

HUD serves the most disadvantaged and disenfranchised communities in the United States. This responsibility is shared among other federal, state, and local agencies, but because of its relationships with local communities and community groups, HUD has a unique ability to introduce local priorities into national dialogs and to provide support and encouragement so that local data meet national standards for inclusion in the national spatial data infrastructure (NSDI).[3] The development of a parcel-level layer for metropolitan areas is particularly important to HUD, to the communities HUD serves, and to national initiatives, including the NSDI and other federal data initiatives.

GIS for Use in Housing and Urban Development

GIS provides a common framework—location—for information from a variety of sources. Because many of HUD's programs have goals that can be identified by location, GIS can be a powerful tool for understanding and managing the results of HUD's efforts (Box ES.1). The modeling and visualization capability of GIS provides a means of testing alternatives and turning data into information, and subsequently into knowledge.

Urban, metropolitan, and state governments throughout the country have already turned to GIS as a means to deal with their own burgeoning demands for more effective and efficient service. Local governments and agencies will want to overlay and analyze the various HUD-developed indices, maps of HUD properties, Section-8 vouchers, etc. on top of other data sources such as base maps, aerial photography, and local analyses and forecasts used daily by a broad assortment of agencies for many urban planning, management, and service delivery purposes. This report provides HUD with direction in their effort to use GIS to better manage relevant data, address their clients' needs, and understand the geographic processes underlying trends in housing markets and the evolution of urban issues in the United States.

HUD's current GIS programs include software-based initiatives (Community 20/20, E [environmental]-MAPS, R [research]-MAPS, and E [enterprise] GIS); research initiatives (programs in the colonias [informal settlements along the U.S.-Mexico border] and Global Urban Indicators); and GIS efforts in HUD's 81 field offices across the nation. These programs are ongoing and evolving. Impediments to the success of these initiatives include lack of clear program ownership leading to discontinuity and inadequate data maintenance; lack of technical input into program design; failure to assess user needs and requirements; data dissemination via CD-ROM rather the Internet; lack of analytic capability in software-based systems, non-integration of relevant datasets, e.g., from other federal agencies or from local sources; and lack of technical and analytic capacity in local HUD staff and recipients of HUD grants.

---

[3]The NSDI is defined as the technologies, policies, and people necessary to promote sharing of geospatial data throughout all levels of government, the private and non-profit sectors, and the academic community <http://fgdc.er.usgs.gov/nsdi/nsdi.html>.

BOX ES.1
Benefits of GIS

Some of the major capabilities of GIS that relate to HUD's mission are described below:

1. GIS provides the **platform for the development of place-based data systems** for measuring the effect of federally-supported housing programs and supporting housing policy decision making. Up-to-date, accurate information is needed to analyze issues and trends, to examine the effect of programs, and to support nationwide analysis.
2. GIS provides the **platform to conduct spatial analysis research** to support policy making and impact assessment. Coupled with the growing availability of spatial analytical tools, GIS permits advanced spatial queries to inform policy making (e.g., "Show me all the housing units with children within 5 miles of a toxic waste site").
3. GIS provides a **platform for collaboration among researchers, practitioners, and policy makers**. GIS is a powerful visualization and communication tool that presents data in a map-like form that people can relate to and offers opportunity for collaborative work on interdisciplinary housing policy questions.
4. GIS provides the **technology to develop Internet-based tools** to support housing decisions for low-income households. Information tools are currently available to higher income households. For example "realtor.com" provides detailed property and neighborhood information for houses available for sale in the private market. Johnson (2002) describes a prototype Internet-based GIS program designed to allow Section 8 participants to identify preferred communities.

SOURCE: Ayse Can Talen, Fannie Mae, personal communication, February 21, 2002.

In 2000, PD&R began to provide the technical expertise, research, and analysis for the development of the EGIS while the office of the Chief Information Officer continues to provide the infrastructure, an enterprise view of the data, and software for data dissemination. This report is intended to provide advice to PD&R and to provide strategic direction for HUD in the agency's use of GIS.

NRC COMMITTEE CHARGE

In 2000, HUD asked the National Academies to evaluate their evolving GIS programs. These programs identify and standardize processes, formats, and specifications for Internet-based, interactive place-based inquiry applications using spatial data from HUD's databases and other federal agency and local agency data sets. To provide perspective and guidance to the PD&R about current and future GIS research and applications and to provide strategic direction for the agency, the committee was asked to pay particular attention to PD&R's specific mission goals including: ensuring the availability and accuracy of essential data on housing market conditions and trends; disseminating this information to the public; conducting research to expand the knowledge base needed for improved policy and practice nationwide; and working through interagency groups to achieve consensus on housing and urban issues.

The staff and committee met with HUD staff and organized a workshop at the National Academies (Appendix B) to discuss a range of housing and urban issues. Workshop participants came from a variety of backgrounds (Appendix A) including federal agencies, state agencies, non-profit public policy groups at the national and local level, universities, and other research organizations. Topics included GIS initiatives at federal and state levels including HUD's GIS programs, spatial analysis of neighborhood change, community-building using GIS, experience from HUD's field offices, regional-scale analysis, and data applications and interoperability. The varied perspectives on data and research needs in urban and community planning informed the committee's deliberations and enriched the report. Based on these discussions and presentations, the committee concluded that a forward-looking approach to GIS research and applications at HUD was more appropriate than a critical analysis of current GIS efforts many of which are ongoing and evolving.

This report takes a regional or metropolitan area[4] approach to housing and urban issues. The alternative is to play what Rusk (1999) calls the "inside game," in which solutions to urban and neighborhood ills are sought in places where they arise rather than in regional processes that have local influence. The conclusions and recommendations of this study provide future direction for HUD and also have relevance for HUD's partners in urban and community planning and decision making at local, regional, and national

---

[4]A metropolitan area is a core area containing a large population nucleus, together with adjacent communities having a high degree of economic and social integration with that core (<http://www.census.gov/population/www/estimates/aboutmetro.html>). For a discussion, see Rusk (1999).

levels in the United States. Although the committee recognizes that expertise in geographic research and analysis is necessary for the implementation of GIS to HUD's mission, addressing workforce and organizational issues related to GIS application at HUD is not within the charge to the committee. Similarly, although the committee recognizes that data initiatives are costly and time-intensive, budgetary considerations are not detailed in this report.

CONCLUSIONS AND RECOMMENDATIONS

Ensuring the Availability and Accuracy of Essential Data

HUD is responsible for providing data for housing and urban decision making nationwide. Steps to ensure the availability and accuracy of essential data include meeting Federal Geographic Data Committee (FGDC)[5] data standards in all operations, taking steps to improve existing data, and creating an internal spatial data infrastructure (SDI). An internal SDI will permit the integration of local data from HUD's partners and support an appropriate urban research agenda for the agency. This is tantamount to the creation of urban framework layers for the National Spatial Data Infrastructure (NSDI). HUD is well-suited to be one of the lead federal agencies in creating an urban SDI for the nation.

1. Be fully FGDC Compliant

Recommendation: As a first step, HUD should meet federal data standards in all operations by:

- **Participating fully in the FGDC and other federal initiatives to assure that agency efforts are consistent with the development of the NSDI; and**
- **Supporting its program participants' efforts to provide operational data in FGDC standard format and to make these data available on the Internet along with other HUD data, subject to the limits of confidentiality.**

HUD datasets are collected from a variety of sources. Local, detailed datasets can be valuable to HUD's local constituents but such data are often

---

[5]The FGDC is comprised of 17 federal agencies and is responsible via an Executive Order of the President (Executive Order 12906, 1994) for the creation and maintenance of the NSDI . (<http://fgdc.gov>).

incomplete and insufficiently documented. For optimal use and to comply with federal data policy, all HUD datasets should be accurate, consistent, and complete. Data from different sources often have to be integrated or "cleaned" spatially and thematically[6] to maximize their utility.

2. Improve data quality

Recommendation: As a first step, HUD should improve existing housing and related data. Existing data should be cleaned and checked for accuracy, consistency, and completeness. Data gaps should be identified and filled. HUD should adopt accuracy and documentation standards that build on FGDC data standards.

To fully comply with the FGDC and other federal initiatives, HUD should develop an in-house, integrated data infrastructure. A spatial data infrastructure can: foster agency-wide data coordination, integration, sharing, and analysis; facilitate internal assessment of HUD programs, and analysis and reporting of federal urban investments; and aid in the delivery of services to HUD clients, such as metropolitan and local governments.

3. Create an agency-wide GIS for urban and community planning

Recommendation: HUD should create an internal spatial data infrastructure for an agency-wide GIS to support an appropriate urban research agenda and to integrate locally derived data.

An agency-wide GIS will permit *inter alia* the evaluation and assessment of the following:

- The strength of prior HUD investments;
- The results of HUD investment on the stability of neighborhoods, municipalities, schools, and school districts;
- The educational and economic opportunity in areas of potential HUD investments; and
- Future investment decisions that will foster health, education, economic opportunity, and residential and commercial stability of neighborhoods and regions.

---

[6]Spatially, when different datasets are combined, boundaries and roads may be topologically inconsistent and require matching. Thematically, two datasets may have different attributes or coding necessitating matching or filling in of missing attributes.

HUD could require public housing agencies to georeference their data, and the accuracy of local data geo-referencing can surpass the TIGER/Line files[7] provided by the Bureau of the Census. HUD's 81 field offices nationwide represent a wealth of local relationships. As a result of these relationships, HUD has significant access to local data and a singular ability to mandate national standards for local data.

4. Integrate local data into the agency-wide GIS

Recommendation: HUD should develop mechanisms to accept and integrate relevant locally-derived data and georeference the data for integration in the agency-wide GIS. Specifically,

- **HUD should spatially enable local data by performing address-matching of individual records at the finest scale using geographic coordinates.**
- **HUD should select, tabulate, analyze, and map relevant housing variables through a GIS at multiple relevant geographic scales (census block, block group, and tract; place, county, and metro-politan area).**
- **PD&R should take the lead within HUD in efforts to integrate grantee and other data at different levels: parcel, neighborhood, municipality, school and school district, metropolitan area, state, and national.**

In this way, data can be made available at multiple scales on a broad range of urban topics, including real estate market conditions, neighborhood educational and economic opportunity, crime, local fiscal capacity, tax rate condition, and environmental risk. Data can be disseminated via the Internet, saving HUD data-users the time-intensive work of data integration.

Create an Urban Spatial Data Infrastructure for the NSDI

This report discusses the growing demand for accurate, relevant data on housing and urban issues; the research required to better understand urban and neighborhood dynamics; the information needed to inform urban and housing policy nationwide; and the evolving roles of diverse partners in urban and community planning. With the agency's GIS programs to monitor

---

[7]Topologically Integrated Geographic Encoding and Referencing System (TIGER) (see <http://www.census.gov/geo/www/tiger/index.html>).

modern urban conditions and respond to emerging trends in housing and urban issues, HUD is well-positioned to be one of the lead federal agencies in providing and managing urban framework data layers for the NSDI. Because HUD carries out its work within a constellation of agencies having varying responsibilities for urban and community issues, the USDI should be developed in partnership with those agencies. That vision for the future is encapsulated in the creation of an urban spatial data infrastructure (USDI) that includes parcel-level data and relevant environmental and socioeconomic data.

5. Create a USDI

Recommendation: HUD should promote the development of a parcel-level data layer and other urban framework layers to create a USDI as a component of the NSDI for housing and urban development. The federal government should make available resources commensurate with this task.

Core elements of the USDI can include:

- Public and federally-assisted housing data;
- Tenant and housing characteristics;
- Parcel-level data;
- Locally updated TIGER files;
- Environmental data; and
- Socioeconomic data.

6. Support development of data centers to support the USDI

Recommendation: HUD should encourage and support the development of local, metropolitan, and regional data centers to facilitate local data coordination, use, and training toward the creation of a USDI.

Disseminate Data and Information to the Public

The Internet is a powerful means for disseminating information to the public. People without a home computer may have access to the Internet at local libraries or community centers. HUD data users range from technically sophisticated urban researchers to local advocacy groups who have little experience with spatial data or technologies. These users have a wide range of data needs and research and analytical capabilities, but all clients and all applications require information about data quality and usability including data interoperability.[8]

7. Create web tools

Recommendation: HUD should continue to develop a spectrum of tools to meet users' needs.

- **For users with limited financial or technical resources, HUD should provide web-based mapping of HUD data and related information.**
- **For more advanced applications, HUD should develop tools for flexible querying, extracting and downloading data, including standard file formats for exchanging data.**

HUD is responsible for providing information about local housing conditions and making basic data on urban and housing issues available to the public. A user-friendly, web-based GIS is an efficient means of data dissemination but the development of a well-designed web-based GIS is a long-term process, and user input is critical. Confidentiality concerns inhibit local sharing and public access to data at needed resolutions.

8. Reach out to users

Recommendation: To improve dissemination and promote the use of spatial data, HUD should:

- **Involve users in design of the web-based GIS;**
- **Sponsor conferences and workshops for clients and partners about using spatial data;**
- **Support online groups for HUD spatial database users; and**
- **Produce an Internet newsletter devoted to spatial data and analysis.**

PD&R is well-positioned to:

- **Work with HUD clients and data users to derive the most appropriate GIS designs and to identify needed data and functions.**

---

[8]Examples of exportable files that are interoperable and can be transferred from one type of software to another include shape files, E00 files, d-base, and tab-delimited ASCII.

- Manage data confidentiality. For some sensitive data, PD&R will need to develop a policy on releasing confidential data as well as algorithms to suppress sensitive data to protect privacy.
- Take the lead in establishing a node for housing and related economic and demographic data in the NSDI's National Geospatial Data Clearinghouse.
- Support the functions of an agency-wide enterprise GIS across all relevant HUD units.

GIS has multiple capabilities and each is valuable in its own way, for basic mapping, data handling, spatial analysis, etc. Local governments and community groups are beginning to take advantage of GIS and are developing capacity to do more sophisticated spatial analysis, yet, at present, most local groups use GIS only for visualization. Visualization allows users to view a few variables but does not support analysis of complex issues facing these communities.

9. Promote users' spatial analytic capacity

Recommendation: To help community groups and local governments develop spatial analysis capabilities, HUD should support the development of tools for spatial analysis. PD&R should support the development of online, down-loadable analytical tools that incorporate multivariate techniques.

Expanding the Knowledge Base for Urban Policy and Practice

HUD is a federal agency with a strategic goal to provide a decent, safe, and sanitary home and equal opportunity for every American. Creating a vision for the future of urban America is an appropriate goal for the agency. This implies a broad urban research agenda.

10. Address research priorities

Recommendation: HUD should expand its research portfolio to emphasize the following urban issues:

- **The spatial distribution of poverty in the United States;**
- **The changing demographics of American neighborhoods; and**
- **Market trends that affect the U.S. housing market.**

To help local governments and non-governmental groups develop their policy analysis capabilities, HUD needs internal spatial analysis capabilities and a systematic approach to monitor metropolitan housing market conditions and trends.

11. Support development of data and tools for monitoring housing markets

Recommendation: To monitor and analyze metropolitan housing market conditions and trends, HUD should:

- **Identify and adopt means and formats for routine collection of housing-related data relevant to user needs and agency mission goals at regular intervals, along with development and adoption of a standardized method for data analysis.**
- **Perform research toward the development of spatial analytic tools to address quality-controlled price indices and variations in local context, and for time-series and comparative analyses between and among places.**

An agency-wide GIS can be used to examine urban issues and housing trends across multiple geographic scales from neighborhood to regional and at various levels of spatial resolution in a metropolitan context. HUD can work with internal datasets and with those produced by partners; investigate the spatial structures and social processes at work in a metropolitan context that underpin many community concerns with housing and investment; and engender participation among partners with interests in policy analysis, research, and community building. An appropriate research agenda can promote the effectiveness of HUD programs.

12. Engage in research to promote program effectiveness

Recommendation: HUD should incorporate into their research agenda and prioritize spatial analysis of the following urban issues at the regional and metropolitan-level:

- **Housing market conditions and trends;**
- **Effects of these conditions on HUD program design and implementation;**
- **HUD program effectiveness and impacts on communities;**
- **Interactions among communities in metropolitan areas;**

- **Dynamics of neighborhood change including poverty concentration, racial segregation, and neighborhood effects; and**
- **Housing and labor market interactions including regional and cross-border analyses.**

Develop and Use Partnerships

HUD can use GIS to facilitate the agency's efforts to interact with organizations beyond its institutional boundaries to build vertical and horizontal networks to share data, discuss housing and urban issues, and ultimately create public policy to respond to these issues. HUD's relationships with state and local groups enable the agency to promote public participation in decision making, narrow the divide that prevents disadvantaged communities from participating in urban and housing policy setting, and bring the capabilities of GIS to bear on issues of local and national relevance.

13. Use partnerships to promote inclusion of local data and perspectives

Recommendation: HUD should facilitate the integration of local datasets and the development of mapping applications using the shared data; encourage public participation in the development and use of local datasets; and partner to develop local and in-house GIS capability.

HUD can partner to link the methodological expertise housed in universities and the local expertise of communities and local governments. Local agencies can develop research questions, analyze research results, and develop appropriate local solutions. PD&R could work with local governments and community groups to define new ways for researchers to extend their skills to building local capacity and addressing local needs.

14. Partner with universities

Recommendation: PD&R should build relationships with university and unaffiliated researchers to engender participation of local groups in policy analysis, research, and community building; and to promote the use of advanced spatial analysis in urban housing policy research to address the complexities of modern urban dynamics.

GIS provides HUD with an opportunity to introduce housing and urban issues onto the national agenda. HUD can partner with other federal agencies having responsibility for providing and managing data relevant to urban, community, and housing issues. Transportation, social services, education, and employment are intimately linked to housing and community development.

14. Partner with other federal agencies

Recommendation: PD&R should take the lead within HUD to build interagency relationships with federal data-providing agencies that have responsibilities related to urban and community issues, notably the Department of Transportation, the Department of Health and Human Services, and the Environmental Protection Agency.

SUMMARY

Steps to ensure the availability and accuracy of essential data include meeting FGDC data standards in all operations, improving existing data, and creating an internal spatial data infrastructure. This work is tantamount to the creation of urban framework layers for the NSDI. HUD is well-suited to be one of the lead federal agencies in creating an urban spatial data infrastructure for the nation. To disseminate data and information to the public, HUD should incorporate users' needs, priorities, and abilities in the design of an agency-wide GIS and associated spatial analytical tools, and develop methods of managing data confidentiality concerns.

HUD is also responsible for conducting research to expand the knowledge base needed to improve urban policy and practice nationwide. To this end, HUD should expand their research to prioritize spatial analysis of poverty in the United States, the changing demographics of American neighborhoods, and economic market trends that affect the U.S. housing market. To achieve these goals, HUD should collaborate effectively with local clients to create local data sets, link university researchers with these clients to promote the devolution of analytical capabilities to communities, and partner with other federal agencies having responsibility for urban, community and housing issues in the United States.

The activities described above require trained personnel to lead and manage. To address these recommendations HUD should consider the addition of spatial data development specialists who have expertise in GIS, spatial analysis, geographic research, algorithm development, and spatial data manipulation. HUD faces formidable challenges including the decline of the inner city, entrenched poverty, and homelessness that resist simple solutions, as well as administrative and resource constraints in implementing spatial data initiatives. This report offers HUD a vision of the future of GIS for housing and urban development.

1

Introduction

HISTORY AND ROLE OF THE DEPARTMENT OF HOUSING AND URBAN DEVELOPMENT

The U.S. Department of Housing and Urban Development (HUD) was established as a cabinet-level agency in 1965. Under Title V1 of the Civil Rights Act of 1964, HUD's Office of Fair Housing and Equal Opportunity is responsible for the agency's federally assisted programs, including housing and community economic development, and for enforcement of related civil rights statutes. The Civil Rights Act of 1968 gave HUD responsibility for enforcing civil rights legislation, including broad housing anti-discrimination laws.[1] The establishment of an agency at the cabinet-level to address civil rights, poverty, and the state of American cities reflects the importance of domestic issues in the "Great Society"[2] programs of that era.

HUD has faced many challenges over its 40-year history as American cities changed and national priorities shifted. International crises, such as the Vietnam War and the gas shortages of the mid-1970s, focused attention away from the domestic issues the agency was created to address. With privatization

---

[1] For details, see the Civil Rights Federal Directory (<http://www.usccr.gov/pubs/crd/federal/hud.htm>).

[2] "Great Society" was the term used to describe domestic policies of President Lyndon B. Johnson who proposed a war on poverty as part of the nation's efforts to overcome racial divisions. President Johnson expanded the federal government's role in domestic policy to advance this goal.

of government functions in the early 1980s, market mechanisms rather than government intervention were identified as the best way to solve problems like poverty, lack of affordable housing, and urban decline. Although policy in this era emphasized market-oriented approaches, Congress also passed the Stewart B. McKinney Act of 1987, which enabled HUD to continue playing a role in addressing housing and urban issues.

The Stewart B. McKinney Act of 1987 engaged HUD to help communities deal with homelessness. New responsibilities for the housing needs of Native Americans, and Alaskan Indians came with the 1988 Indian Housing Act. HUD's priorities for increasing home ownership, especially for low-income Americans, were reinforced by the Cranston-Gonzalez National Affordable Housing and Low-Income Housing Preservation and Residential Homeownership Acts of 1990. In 1995, the "Blueprint for Reinvention of HUD" stressed housing reform, adaptation of the Federal Housing Administration, and consolidation of programs into community block grants. Through the Quality Housing and Work Responsibility Act of 1998, Congress approved Public Housing reforms to reduce segregation by race and income, encourage and reward work, bring more working families into public housing, and increase the availability of subsidized housing for poor families.[3] Home-ownership reached a record high in the third quarter of 2000, when 67.7 percent of American families owned their homes (HUD, 2002a).

Despite changes in emphasis and national priorities over HUD's history, the agency's mission has remained constant. HUD's mission is to increase homeownership, support community development, and increase access to affordable housing free from discrimination (HUD, 2002b).

To carry out its mission, HUD has identified six strategic goals (HUD, 2002b):

1. Increase homeownership opportunities.
2. Promote decent affordable housing.
3. Strengthen communities.
4. Ensure equal opportunity in housing.
5. Embrace high standards of ethics, management, and accountability.
6. Promote participation of faith-based and community organizations.

HUD's wealth of relationships with local and community groups and attention to the most disadvantaged communities in the United States provide

---

[3]Poverty measurements are variable. The U.S. Bureau of the Census sets income thresholds that vary with locality and with family size and composition, below which a family is considered poor (<http://www.census.gov/hhes/poverty/povdef.html>). See Citro and Michael, 1995.

the agency with a singular ability to introduce the priorities of local and underrepresented groups into national dialogs. The agency can also advance the inclusion of much needed local, urban data that meets national standards in the national spatial data infrastructure. The Office of Policy Development and Research (PD&R) is at the forefront of HUD's efforts to use geographic data and tools to address issues of urban poverty and community housing needs.

HUD'S OFFICE OF POLICY DEVELOPMENT AND RESEARCH

To carry out its mission, HUD engages in research and develops policies help socially and economically disadvantaged Americans secure adequate housing in safe communities. PD&R contributes to HUD's mission by maintaining current information to monitor housing needs, housing market conditions, and the operation of existing programs; and by conducting research on priority housing and community development issues (HUD, 2002b).

As described below, PD&R has worked toward the creation of an agency-wide geographic information system (GIS) to bring together the data that HUD and its clients and partners need to address issues of housing and urban development and to promote citizens' participation in related decisions. HUD asked the National Research Council to evaluate the agency's GIS programs with particular attention to PD&R. The committee addressed the potential of GIS and spatial analysis for supporting PD&R's goals including: ensuring the availability and accuracy of essential data on housing market conditions and trends; disseminating this information to the public; conducting research to expand the knowledge base needed for improved policy and practice nationwide; and working through interagency groups to achieve consensus on housing and urban issues.

GEOGRAPHIC DATA FOR POLICY AND RESEARCH AT HUD

Geographic information is a subset of the overall information base at HUD (Box 1.1). Public sector agencies such as state, local, and federal government need geographic information to carry out missions, such as resource conservation, species protection, infrastructure planning and maintenance, surveying and mapping, and land-use analysis. HUD needs geographic information to carry out its mission to increase homeownership, to support community development, and to increase access to affordable housing free from discrimination. To meet these goals, HUD must know the location of

Box 1.1
Information Systems and GIS

Information systems and GIS have much in common because both use a computerized system for collecting, storing, analyzing and disseminating information. An important difference is GIS' capability for cataloging spatially referenced objects and their attributes within the context of a map. GIS can also be used to perform quantitative analysis based on geographical principles, such as location in space, flows, or vertical and horizontal relationships (Obermeyer and Pinto, 1994).

Research issues at the intersection of computer science and geographic information sciences include database management systems, data mining, human-computer interactions (including graphics and visualization), transmission of vector data via the Internet, and development of algorithms and data structures (NRC, 2002a,b).

The management of information systems across an organization and the development of an agency's enterprise GIS demand the formulation of long-term objectives and plans at a high management level; as well as operational decisions, goal setting, resource allocation, and staff development at the middle management level (Obermeyer and Pinto, 1994). See Figure 1.1.

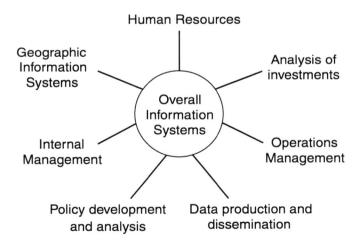

FIGURE 1.1 This Venn figure demonstrates the relationship between Information Systems and GIS. SOURCE: Adapted from Obermeyer and Pinto, 1994.

existing housing and the families who are eligible for that housing. If communities are to develop, jobs and services must be accessible, and GIS can help determine and show the spatial relationships among these amenities. To understand the housing market in a neighborhood, HUD must have information about economic and market trends in the surrounding city and region.

HUD's current GIS programs include software-based initiatives (Community 20/20, E[environmental]-MAPS, R[research]-MAPS and E [enterprise] GIS), research initiatives (programs in the Colonias and Global Urban Indicators), and GIS efforts in HUD field offices. These programs are discussed in more detail where relevant throughout the report.

As in many federal agencies, the proportion of political appointees at HUD creates frequent turnover and may contribute to reduction in continuity of GIS initiatives. Community 20/20 (Box 2.2), initiated in 1994, suffered from lack of clear program ownership leading to problems with data maintenance and updating, and contributing to the lack of assurance of data quality and comparability across HUD units. Community 20/20 software was not made available online, which impeded widespread use. No assessment of users' needs or requirements was carried out and there was little input into the initiative from technical units at HUD.

E–MAPS (Box 2.3), a partnership begun in 2000 between HUD and the U.S. Environmental Protection Agency to make environmental information available at the neighborhood level, provides spatially referenced data that is essentially descriptive in nature. Information is made available free over the Internet. No analytic capabilities are included in E-MAPS and no new data can be overlaid on existing information. R–MAPS (Box 2.4) marked an improvement in spatial analytic capabilities and was a prototype for the current web-based interactive mapping system in HUD's enterprise GIS (EGIS). Datasets from other federal agencies were not included in R-MAPS.

HUD's EGIS (Box 2.5), developed in partnership with Environmental Systems Research Institute (ESRI), sets out to correct these problems. EGIS is housed in the Office of the Chief Information Officer (CIO) at HUD. In 2000, PD&R began to provide the technical expertise, research, and analysis for the development of the enterprise GIS while the CIO continues to provide the infrastructure, an enterprise view of the data, and software for data dissemination. With EGIS, data and user-friendly tools will be provided via the Internet. The addition of two sets of census data to the EGIS will facilitate assessment of data trends. The EGIS initiative also indicates the agency's intention to cooperate with other federal agencies that produce data related to HUD's mission but there is no indication that HUD intends to integrate local datasets into the EGIS or make local datasets available via the Internet as part of this initiative. The EGIS is intended to provide an enterprise view of HUD data, spatial analytic capacity, and to address issues of data accuracy and completeness.[4]

The internal use of GIS at HUD within its regional offices varies considerably. Recipients of HUD grants tend to use GIS to identify the spatial distribution of their programs; to visually display allocation of resources for political and educational purposes; and for advocacy related to programmatic directions. Currently, HUD's field offices often lack both adequate data and staff who are proficient in GIS and spatial analysis. Due to limited understanding of spatial analysis, comparative spatial statistics, and housing indicator development, most of the GIS efforts in HUD's field offices go no further than point and thematic mapping.

This report focuses on PD&R's role in providing technical and research advice and information across HUD's program offices and in support of the development of the EGIS. PD&R performs research and analysis of spatial data for all of HUD's program offices. The major components of HUD's research programs using GIS are the following:

- Economic Affairs: analysis of mortgage trends, mortgage distribution, home ownership, and general demographic information for determination of fair market rent.[5]
- American Housing Survey: provision of national and metropolitan estimates of housing conditions and zoned data.
- Policy Development: research leading to improved policy in support of mission goals.
- Research and Evaluation: post-program evaluation for fair housing programs, assistance programs, and mortgage market assessments; and mapping of racial and economic segregation, and mapping.

Among the data-collection activities HUD funds are the American Housing Survey, carried out by the U.S. Bureau of the Census; Fannie Mae and Freddie Mac home mortgage data collections; and the *State of the Cities Report*[6] from which HUD creates value-added products on employment, education, and other indicators of urban quality-of-life. HUD has access to standard U.S. Census data, 1990 special tabulation of non-standard data,

---

[4]Current information about the EGIS is available at <http://hud.esri.com>.

[5]The current definition of fair market rent is the 40th percentile rent—the dollar amount below which 40 percent of standard quality rental housing units rent. National Low Income Housing Coalition (<http://www.nlihc.org/oor2000/appendix.htm>).

[6]<http://www.huduser.org/publications/polleg/tsoc99/contents.html>.

statistics for cities (entitlement communities), Federal Housing Administration (FHA) mortgage data, fair housing data, and Fannie Mae and Freddie Mac data—all at the census-tract level.[7]

GIS AND A CHANGING AMERICA

Trends over the last 20 years have transformed the organization and practice of everyday life in America. High-speed information and communication technologies and related infrastructures link people and organizations in what Manuel Castells (1996) calls "the network society." The availability of information and networks for sharing it has transformed economic practices ranging from manufacturing to finance to marketing (creating, for example, the "e-economy"), and our social lives (from daily household tasks to the creation of new national and international communities). At the heart of the Information Society are digital technologies for the collection, storage, retrieval, analysis, and mapping of information. One of these information systems—GIS—deals with the handling, analysis, and mapping of spatially referenced data.[8] GIS has transformed the practices of research and application in dealing with environmental, land use, resource management, and urban and regional issues in the United States.

GIS is a computer tool for understanding where things are located on the surface of the Earth. More than simply drawing maps, GIS allows analysts and citizens to answer questions such as: "Where is assisted housing located in my neighborhood?" "Which environmental hazards are within 5 miles of where I live?" "How can I get from where I live to where I can work?" "What is the best location for a day-care facility in my community?" GIS does this by permitting the integration and analysis of different kinds of data (e.g., environmental data and demographic data), and the display of the results in a visually attractive, and graphically explanatory way.

Why Is GIS Important to HUD and PD&R?

GIS can aid in public-policy decisions for more effective allocation of resources for community and economic development, for better-managed community planning and growth, as well as for the efficient delivery and use

---

[7]A census tract is an areal unit used in collecting and reporting census data (NRC, 1998).
[8]Spatially or geographically referenced data are data with known latitude, longitude, and elevation, or other horizontal and vertical coordinates.

of public services. It provides a common framework—location—for information from a variety of sources. Because many of HUD's programs have goals that can be identified by location, GIS can be a powerful tool for understanding and managing the results of HUD's efforts (Box 1.1). Moreover, the modeling and visualization capability of GIS provides a means of testing alternatives and turning data into information, and subsequently into knowledge. GIS is a tool for data management and spatial analysis but the information derived from GIS is only as accurate as the data that went into the system and as relevant as the questions posed.

The use of GIS has increased markedly over the last 10 years in the United States and worldwide. Reasons include increased capabilities of computers generally, and GIS specifically; decreased costs of computer technology; and development of user-friendly, web-based interfaces (Obermeyer and Pinto, 1994). The evolution of software and data standards, supported in part by efforts of the Federal Geographic Data Committee (FGDC), has spurred the application of geographic information and the diffusion of GIS by making data more available and more readily usable for a range of applications. Growing familiarity in the general population with spatial technologies such as GIS and the global positioning system (GPS) has also increased the demand for geographic information and tools.

The potential of GIS to handle large quantities of spatial data makes the technology valuable for a variety of management and planning decisions about land use, community development, and resource allocation (NRC, 1999). GIS can contribute to the HUD's mission in numerous ways as described in Box 1.2.

Urban, metropolitan, and state governments throughout the country have already turned to GIS as a means to deal with their own burgeoning demands for more effective and efficient service. Public and private services and utilities have been mapped digitally. Emergency services, such as 911, make regular use of GIS systems to enhance their performance and response times. Urban authorities are using GIS to study the rapid changes in the social and spatial make-up of neighborhoods. HUD is using GIS to better manage spatial data and better understand the spatial processes underlying housing market trends and the evolution of urban issues in the United States.

HUD AND THE EVOLUTION OF URBAN ISSUES

The nature of the city and the metropolitan region has been drastically transformed by urban expansion and related economic decentralization. In many cases, a deepening of socioeconomic differences within the city has resulted. Poverty in the city is a product of regional growth and the transfer of the tax base and regional expenditures from poor urban areas to thriving suburban areas (Orfield, 1997). The flight of the middle class and of the jobs it once sustained in the city has greatly complicated HUD's central mission of providing adequate housing and alleviating poverty. Potential solutions are neither simple nor subject to consensus by all concerned.

Box 1.2
Benefits of GIS

Some of the major capabilities of GIS that relate to HUD's mission are described below:

1. GIS provides the **platform for the development of place-based data systems** to measure the impact of federally supported housing programs and support housing policy decision-making, Up-to-date, accurate information is needed for analyzing issues and trends, for examining the impact of programs, and to support nation-wide analysis.
2. GIS provides the **platform to conduct spatial analysis research** to support policy making and impact assessment. Coupled with the growing availability of spatial analytical tools, GIS permits advanced spatial queries to inform policy making (e.g., "Show me all the housing units with children within 5 miles of a toxic waste site").
3. GIS provides a **platform for collaboration among researchers, practitioners, and policy makers.** GIS is a powerful visualization and communication tool that presents data in a map-like form that people can relate to and offers opportunity for collaborative work on interdisciplinary housing policy questions.
4. GIS provides the **technology to develop Internet-based tools** to support housing decisions for low-income households. Information tools are currently available to higher income households. For example "realtor.com" provides detailed property and neighborhood information for houses available for sale in the private market. Johnson (2002) describes a prototype Internet-based GIS program designed to allow Section 8 participants to identify preferred communities.

Source: Ayse Can Talen, Fannie Mae, personal communication, February 21, 2002.

During the 1990s, HUD developed an in-house GIS and expanded access to its web-based system showing loan-performance data. Then-Secretary Cuomo described the program as "another step in our plan to protect consumers and communities from predatory and abusive lending practices" (Hasson, 2000). GIS offered an approach to address the dynamics of urban poverty, housing provision, and regional growth at multiple scales including the local and the regional. Increased attention to the threat of international terrorism since September 11, 2001, has increased pressure for the integration of information systems and for speedy access to useful and reliable information. It is within this framework of evolving technology, urban-regional dynamics, and national needs that HUD seeks an assessment of its use of GIS.

Through PD&R, HUD has responded to the increased availability of geographic information and technologies with a number of software- and research-based initiatives that use GIS. These efforts have focused on long-standing agency priorities including affordable housing and home ownership, as well as community empowerment through access to information and decision-making tools. Descriptions of these efforts are summarized in textboxes throughout Chapter 2.

Providing affordable housing is among HUD's principal and long-standing goals. An evaluation of HUD-assisted public housing programs by the National Academy of Public Administration (NAPA) recommends that HUD improve its organizational culture to enhance collaboration with its housing partners. Specifically, the NAPA report suggests that HUD make longer-term systemic improvements to increase flexibility, reduce administrative and data burdens, and take advantage of opportunities for greater use of outcome-oriented techniques to enhance housing quality (McDowell, 2001). Although this report does not discuss organizational issues, the contributions of an agency-wide GIS to HUD's internal program assessment and monitoring is described in Chapter 2.

The agency routinely updates its information systems but HUD's current software systems (that is, Public and Indian Housing Information Center System, the Real Estate Management System, the Residential Assessment Subsystem, HUD's Central Accounting and Program System, and the Empowerment Information System) are not adequate to the task at hand. A report by the General Accounting Office (GAO) finds that HUD's current information systems do not meet the information needs of management and staff and do not provide support for needed programs and operations due to weaknesses in developing requirements, project management, contract tracking, and software evaluation (GAO, 2001). HUD is aware of these weaknesses and the need for integrated information across the agency—this study was requested as part of HUD's efforts to address these issues.

High-quality and reliable information systems, including GIS, are needed to support HUD's goals of making housing affordable, revitalizing communities, and encouraging home ownership. These information systems are also needed to support HUD's internal management, financial and administrative programs, and research agenda. HUD functions in a complex institutional network of individuals, government agencies at levels from local to federal, non-governmental organizations, private companies, and its own dispersed regional and field offices. Good communication and the availability and exchange of essential data on housing and urban development are necessary for this network to function.

STUDY AND REPORT

The report evaluates HUD's evolving GIS programs. These GIS programs identify and standardize processes, formats, and specifications for Internet-based, interactive place-based inquiry applications using spatial data from HUD's databases and other federal agency and local agency datasets. Particular attention is paid PD&R's specific mission goals including: ensuring the availability and accuracy of essential data on housing market conditions and trends; disseminating this information to the public; conducting research to expand the knowledge base needed for improved policy and practice nationwide; and working through interagency groups to achieve consensus on housing and urban issues. At the request of HUD, this study was undertaken to provide perspective and guidance to the PD&R about current and future GIS research and applications, and to provide strategic direction for the agency.

The committee chose to address urban and housing issues at the regional and metropolitan level[9] because that is the best way to identify and address the patterns and processes that determine poverty in the United States whether it occurs in the city or in the countryside. The alternative is to play what Rusk (1999) calls the "insider game" in which solutions to urban and neighborhood ills are sought in the local places where they arise rather than in regional processes that have local influence. Poverty in the United States is

---

[9]A metropolitan area is a core area containing a large population nucleus, together with adjacent communities having a high degree of economic and social integration with that core (<http://www.census.gov/population/www/estimates/aboutmetro.html>). For a discussion, see Rusk (1999).

largely an urban phenomenon,[10] but it is the product of social and economic processes working at the metropolitan and regional level (Brockerhoff, 2000; Lichter and Crowley, 2002; Orfield, 1997). These processes include middle-class flight, dis-investment in inner city areas, and poverty concentration. Moreover, they involve factors such as access to transportation, jobs, and social services that are also effectively addressed at the regional level. For example, research shows that poverty is less concentrated if low-income housing is provided in a desirable location, if training and job placement programs are provided for low-income households, and if a critical mass of non-subsidized units is maintained in the area especially for an intermediate income range (Brophy and Smith, 1997).

The charge to the committee and the principal aim of this report is to provide perspective and guidance to the PD&R about current and future GIS research and applications and to provide strategic direction for HUD. Since HUD does not operate in isolation, the committee also considered the network of people who use HUD's data for policy and research purposes. This group includes professors and students at colleges and universities, policy makers and analysts working for local governments, HUD program managers and research scientists, and neighborhood leaders and residents employed by community-based organizations. The committee heard presentations from and interviewed representatives from public, non-profit, and private sector groups at the national, state, and local levels (Appendices B and C). In addition, committee deliberations were informed by contributions from a wide variety of participants (Appendix A) in a workshop held at the National Academies on April 25-27, 2001 (Appendix B).

Chapter 2 discusses HUD's responsibility for providing accurate and relevant data on urban and community issues and identifies the development of national urban framework data layers as an appropriate goal for the agency. Chapter 3 examines programs and tools for disseminating information to a range of clients and partners in urban and housing policy arenas. Chapter 4 outlines a research agenda for PD&R that can support HUD's mission. Chapter 5 discusses the role of partnerships for carrying out research and providing information for urban and community planning in the United States and abroad. The committee's conclusions and recommendations to HUD for its urban and housing data and research agenda are detailed throughout the chapters.

---

[10]In 1998, in the United States, urban poverty was 12.3 percent compared with 14.4 percent for rural areas, but because the population is 75 percent urban, there are more urban than rural poor in the United States (Brockerhoff, 2000).

2

Ensuring the Availability, Accuracy, and Relevance of Urban and Housing Data

Because of HUD's relationships with local communities and community groups, HUD can introduce local priorities into national dialogs, and can support and encourage providers of local data to meet national standards for inclusion in the National Spatial Data Infrastructure (NSDI).[1] The development of a parcel-level layer of information for metropolitan areas is particularly important to HUD, to the communities HUD serves, and to the NSDI and other federal data initiatives. This chapter discusses the challenge that HUD faces in its mission to provide urban and housing data for the nation, and suggests methods to ensure the accuracy and relevance of these data. The final section describes how fully responding to this spatial data challenge is tantamount to creating urban framework data layers for the NSDI.

THE SPATIAL DATA CHALLENGE

Access to reliable spatial data is essential for research, analysis, and policy development on numerous urban and housing issues. Such data are fundamental building blocks for all agencies and organizations that have

---

[1]The NSDI is defined as the technologies, policies, and people necessary to promote sharing of geospatial data throughout all levels of government, the private and non-profit sectors, and the academic community (<http://fgdc.er.usgs.gov/nsdi/nsdi.html>).

resource management and allocation mandates.[2] Reliable spatial data and technologies are needed to monitor and manage urban growth; maximize social, environmental, and economic well-being; and achieve important long-term goals related to quality of life. Without relevant and accurate spatial data at the base, GIS and related technologies (e.g., the global positioning system [GPS], remote sensing, computer mapping, and spatial analysis) are useless. Furthermore, tools for analysis and decision support are required for the application of geographic data to real-world issues.

Public and private institutions are making resources available for long-term decisions about the collection, management, and use of spatial data (NRC, 1997). The Federal Geographic Data Committee (FGDC), representing 17 federal agencies, coordinates the development of the NSDI (Box 2.1). The NSDI encompasses policies, standards, and procedures for organizations to cooperatively produce and share geographic data and information. It is being developed in cooperation with organizations from state, local, and tribal governments; the academic community; and the private sector.

In the United States, geographic data collection is a multibillion-dollar business. In 1993, the U.S. Office of Management and Budget conducted a survey and found that total annual expenditure in federal agencies alone was close to $4 billion.[3] Another estimate places total annual revenues from GIS hardware, software, and data sales at $7 billion in 1999, with the GIS data industry being the most significant sub-sector (Longley et al., 2001).

Developing, maintaining, and disseminating reliable spatial data has been a major challenge to many organizations. Without a coordinated effort, duplicate data for the same locality or region could be collected by multiple organizations using various definitions of time, at different spatial scales, and with varying degrees of accuracy. In the United States, President Clinton's 1994 Executive Order 12906 established the NSDI and set a significant mile-stone in coordinating spatial data development (see Box 2.1).

The NSDI is evolving, and its strategic goals have been redefined to reflect various stakeholders' input and current trends. Of NSDI's many activities, three have significant implications and direct applicability to the development of spatial data and GIS technology at HUD. They are the development of data standards, the development of framework data and the geospatial data clearinghouse, and the establishment of partnerships with state, local, private sectors, and local communities.

---

[2] See NRC (2002c) for discussion of federal data collection and dissemination.
[3] Source: FGDC web page, <http://www.fgdc.gov>.

BOX 2.1
The National Spatial Data Infrastructure

Created partly because of the recommendation of a National Research Council report (NRC, 1993), the NSDI refers to the technologies, policies, and people necessary to promote sharing of geospatial data throughout all levels of government, non-profit organizations, the private sector, and the academic community. The goals of the NSDI are to reduce duplication of effort among agencies; improve quality and reduce costs related to geographic information; make geographic data more accessible to the public; increase the benefits of using available data; and establish key partnerships with states, counties, cities, tribal nations, academia, and the private sector to increase data availability.

TABLE 2.1   Various Responsibilities for Data Layers of the NSDI

Subcommittees	Federal Agency Chair
Base Cartographic Data	U.S. Geological Survey
Cadastral	Bureau of Land Management
Cultural and Demographic	Bureau of Census
Federal Geodetic Control	NOAA's National Geodetic Survey
Geologic	U.S. Geological Survey
Ground Transportation	Bureau of Transportation Statistics
International Boundaries & Sovereignty	Department of State, Office of Geographer
Marine and Costal Spatial Data	NOAA's Coastal Services Center
Soils	USDA's National Research Conservation Service
Spatial Climate	USDA's National Water and Climate Center
Spatial Water Data	U.S. Geological Survey
Vegetation	USDA's U.S. Forest Service
Wetlands	U.S. Fish and Wildlife Service

SOURCE: Adapted from FGDC Chart of Partner Responsibilities.

To enable data sharing among different data-producing units, some basic information about the data, that is, the metadata[4] must be provided. FGDC data standards give terminology and definitions for the documentation of digital geospatial data. Included in the data standard are information on data availability for a geographic location, their fitness for intended use, how to access the data, and how to process and use the data. All federal agencies are required to work with the FGDC. Each agency is responsible for:

- Cooperating as requested in the development of appropriate coordinating mechanisms;
- Supplying necessary information to the interagency coordinating committee concerning its surveying, mapping, and related spatial data requirements, programs, activities, and products; and
- Conducting its surveying, mapping, related spatial data gathering and product distribution activities in a manner that provides effective government-wide coordination and efficient, economical service to the general public.[5]

The concept of a framework and a clearinghouse ensure that data now being developed are made available to many users. The National Geospatial Data Framework is designed to be a collaborative effort. It currently contains seven data themes, including transportation, hydrology (rivers and lakes), geodetic control, digital imagery, government boundaries, elevation and bathymetry, and land ownership (Figure 2.1).

The framework data represent the best available data that are certified, standardized, and described according to a common standard. They provide a foundation on which organizations can build by adding their own data. Moreover, through the National Geospatial Data Clearinghouse (a distributed, electronically connected network of geospatial data producers, managers, and users), additional data can be accessed that is layered on top of the framework data.[6] The appropriateness of the various spatial data can then be determined from their metadata descriptions, which are required for all datasets included in the Clearinghouse.

---

[4]Spatially enabled data include horizontal and vertical coordinates and meta-data, or "data about data", describing the content, quality, condition, and other characteristics of the data.

[5]1990 Office of Management and Budget (OMB) A-16.

[6]See <http://nsdi.usgs.gov/> for a description of the National Spatial Data Clearinghouse.

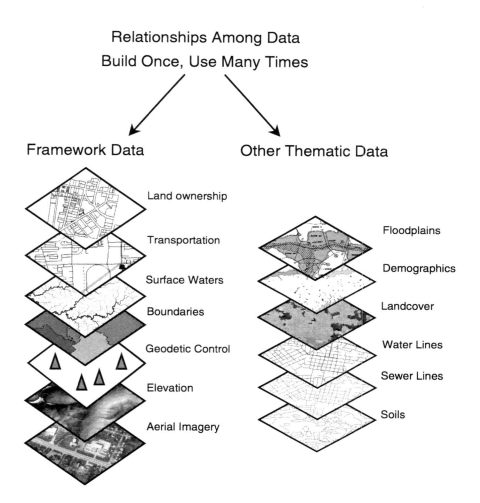

Relationships Among Data
Build Once, Use Many Times

Framework Data

- Land ownership
- Transportation
- Surface Waters
- Boundaries
- Geodetic Control
- Elevation
- Aerial Imagery

Other Thematic Data

- Floodplains
- Demographics
- Landcover
- Water Lines
- Sewer Lines
- Soils

FIGURE 2.1   Once framework and thematic data foundation layers are built, the data can be used for many applications. SOURCE: NSGIC and FGDC, nd.

The success of data development activities depends on partnerships with state and local governmental and non-governmental organizations to continue building data and enriching the NSDI Clearinghouse. The NSDI is characterized by strong partnerships and collaborations, and strategic goals of increasing participation through education and outreach to derive common solutions for discovery, access, and use of geospatial data in response to the needs of diverse communities. Attention is given to community-based approaches to developing and maintaining common collections of geospatial data for sound decision making.[7]

In the last decade, HUD has launched other initiatives that are closely related to the NSDI and directly relevant to the development of spatial data and the use of GIS. HUD launched Community 20/20 in 1994 (Box 2.2), and also partnered with EPA in a web-based initiative called E-MAPS (Box 2.3). Subsequently, HUD introduced another set of geospatial data CD-ROMs called R-MAPS (Box 2.4). The agency's Enterprise GIS is currently under development (Box 2.5).

HUD has initiated several research-based uses of GIS at PD&R. Among the first were those associated with Community 20/20 (See Box 2.2), and examples are reflected in *Mapping Your Community* (HUD, 1998). Several other targeted applications emerged, including an investigation of the value of computer mapping to questions about mortgage lending (Wyly and Holloway, 2002); an evaluation of evidence for segregation and discrimination (NRC, 2002d); and an assessment of patterns of crime around public housing (HUD, 1999). These projects were part of a broader consideration of the use of spatial data analysis and GIS at HUD and related institutional needs. Ongoing GIS research at HUD includes the U.S.–Mexico Border initiatives, the Global Urban Indicators project, and community-based GIS efforts. These efforts are discussed in Chapter 5 (see Boxes 5.1 and 5.2).

Large-scale and long-term research projects such as these require interdisciplinary expertise and good datasets to address complex environmental and societal problems. GIS as an enabling technology can play an important role in supporting large-scale, data-intensive research. Research initiatives such as National Science Foundation's Information Technology Research, and Biocomplexity in the Environment initiatives, NASA's Earth Science Enterprise and its program to detect land use and land cover change, and EPA's regional assessment initiative are examples of such recent research trends.

The Digital Earth concept,[8] embraced by the research community in the late 1990s, refers to a multi-resolution, three-dimensional representation of the planet where vast quantities of geospatial data are embedded. Promoted by Vice President Gore in 1998, the Digital Earth concept provides a vision whereby geospatial data, methods, and analyses are combined so that important societal issues—such as crime, biodiversity, global change, and food security—can be tackled in a more timely and efficient manner.

---

[7] See <http://www.fgdc.gov> for details.
[8] See <http://www.digitalearth.gov/> for details.

BOX 2.2
Community 20/20

In 1994, HUD developed a geographic and statistical data resource called Community 20/20. The product aimed to provide multi-faceted planning, mapping, and communication capabilities to HUD data users. The software, which received the Ford Foundation and Harvard University's Kennedy School Innovation in American Government Award in 1996, was sold on CDs but provided free of charge to fair housing centers nationwide. HUD sold 3,500 copies of the CD and allocated 2,000 free copies (Dick Burke, U.S. HUD, personal communication, 2002).

The purpose of this GIS product was to allow citizens to see the investments of 60 active HUD grant programs. Demographic data, encompassing 600 data items from the 1990 Census, and federal Empowerment Zone and Enterprise Communities activities were linked through this package (Thompson and Sherwood, 1999). Socioeconomic data from the U.S. Census and other sources included data on births, deaths, crimes, school performance, housing code violations, property values, and toxic emissions. These data could be coupled with information about types and locations of HUD-funded projects. The software was specifically designed to support the consolidated planning activities that HUD requires of local governments.

Community 20/20 was intended to provide diverse groups (community-based and non-government organizations, state and local governments, and housing authorities) with the capability to plan and design housing and urban development projects. To promote the use of these tools for community development, HUD published Mapping Your Community: Using Geographic Information to Strengthen Community Initiatives (1998). These products are used primarily in connection with specific HUD proposals and projects.

The Digital Earth concept extends the spatial data challenge further into an international realm. HUD, as a national agency, could play an important role in global housing and habitat studies. The Global Spatial Data Infrastructure (GSDI), established in 1996 by a number of nations and organizations, recognizes the importance of geographic data and supports ready global access to geographic information.[9] Among its goals, the GSDI aims to promote awareness and implementation of complementary policies, common standards, and effective mechanisms for the development and availability of interoperable

---

[9] See <http://www.gsdi.org> for details.

digital geographic data and technologies to support decision making at all scales for multiple purposes.

Finally, the popularization of the Internet in the last few years has revolutionized the concept and practice of NSDI and the related initiatives discussed above. The Internet allows cost-effective, user-oriented dissemination of data and information. It also enables user input and encourages online collaboration. On the other hand, Internet dissemination exacerbates some existing data concerns, such as confidentiality, privacy, and control.

As administrative data (e.g., local data on income tax, employment, and public assistance) are computerized and geographically referenced, and as spatially disaggregated data become standardized and available, privacy concerns will mount. Concerns include the possibility that the data and the developed indicators might be used for red-lining or otherwise stigmatizing troubled neighborhoods. These questions are growing in complexity and importance and deserve increased attention by federal agencies such as HUD. New technologies provide a means to address the need for both detailed, local data and for privacy and confidentiality concerns by allowing users to analyze disaggregated data without giving them case-by-case access. These are major challenges for agencies; nevertheless, full and effective participation in mandatory federal data initiatives demands attention to such questions. Efforts that HUD undertakes to meet FGDC standards will also benefit HUD's internal efforts to collect, use, and disseminate information on urban and housing issues.

Conclusion: To participate fully with the FGDC and other federal data initiatives, HUD should develop an in-house, integrated data infrastructure. To provide reliable data and be consistent with the NSDI, data should be accurately described and assigned spatial definition (geo-referenced) according to the standards of the FGDC.

Recommendation: As a first step, HUD should meet federal data standards in all operations by:

- **Participating fully in the FGDC and other federal initiatives to ensure that agency efforts are consistent with the development of the NSDI; and**
- **Supporting its program participants' efforts to provide operational data in FGDC standard format and make these data available on the Internet along with other HUD data, subject to the limits of confidentiality.**

BOX 2.3
E-MAPS

In September 2000, HUD launched a partnered effort with the U.S. Environmental Protection Agency (EPA), linking data on HUD-funded activity in every neighborhood across the country with EPA environmental information. The purpose of Environmental-MAPS or E-MAPS is to provide people with detailed, site-specific information about what the government is doing to protect the environment and to promote community and economic development. The goal is to ensure easy access to data so that communities can engage in informed discussions and make informed decisions about growth and development. Data available through E-MAPS include:

- the location, type and performance of HUD-funded activities;
- site-specific information about all Superfund sites and related laws;
- brownfields data and Brownfields Tax Incentive Zones; and
- other environmental data including air pollution reports, toxic chem.-icals data, hazardous waste business and permitting information, trend analyses of hazardous waste generation, and company waste water discharge information.

Communities interested in redeveloping abandoned or underused industrial sites can use the data to check for contamination and determine what financial resources exist for redevelopment in the area. (HUD, 2000). HUD E-MAPS are intended to enable communities to make informed decisions about new sites for facilities, such as public and assisted housing, and to help communities prioritize the demolition of existing complexes.

E-MAPS can be accessed on-line at: <http://198.102.62.140/emaps/SearchFrame.asp>.

The trend is clear. With the Internet, public demand for information services will increase, and so will participation in community-based activities that use these data. Solutions to many real problems that exist today require teamwork and collaboration. The development of spatial data coupled with GIS technology is necessary for a federal agency such as HUD to continue to function efficiently in this information age and to be responsive to societal needs.

ENSURING DATA ACCURANCY AND RELEVANCE

Current Data Sets at HUD

PD&R maintains a number of housing-related databases. To increase awareness of the availability of these data, PD&R has compiled *The Guide to PD&R Data Sets*, which can be downloaded from the HUD USER web site.[10] The guide describes 13 available housing data resources and provides web links to related documents and datasets. Each dataset is provided with basic information, such as the source, geographic coverage, period covered, web address, background, intended users, and intended use. These important components of metadata will enable researchers to find the data quickly and easily. The following is a brief synopsis of these data sources.

The **Low-Income Housing Tax Credit (LIHTC)** database contains information on over 16,600 projects and nearly 710,000 housing units placed in service nationwide between 1987 and 1998. Geographic data for each housing project include its address, census tract, city, county, metropolitan area, and state. **The Qualified Census Tracts** dataset contains information on tracts that are qualified for the low-income housing tax credit based on the 1990 census data. The dataset covers all of the United States. The **Difficult Development Areas** dataset, also a national dataset, includes information on areas where incomes are substantially lower than housing costs. The data are broken down by state, the metropolitan statistical areas (MSAs), and non-MSAs.

The **American Housing Survey (AHS)** includes two datasets. The national dataset is a nationally drawn sample of 60,700 housing units covering 878 counties and independent cities throughout the United States. The data provide detailed information on housing conditions as well as characteristics of householders, such as apartments, mobile homes, family composition, income, and neighborhood quality. Geographic indicators for each housing unit include the census region and whether it is in a central city, suburb, or non-metropolitan area. AHS's **metropolitan sample** includes some 5,000 housing units from 47 metropolitan areas. The smallest geographic area identified for each unit is the zone. Zones are groups of census tracts where at least 100,000 persons live.

The **Property Owners and Managers Survey (POMS)** was designed to provide information about the cost and availability of rental housing and what motivates owners to rent out their property to tenants. Although a nationwide survey was conducted, the final POMS dataset includes locational

---

[10] Available at <http:// www.huduser.org>.

information of only 438 sampling areas. Geographic indicators of each housing unit include the census region and whether it is in a central city, suburb, or non-metropolitan area. The **State of the Cities Data Systems (SOCDS)** consists of five databases for many metropolitan areas, central cities, and suburban places. These five databases provide information on each locality, its historical census, poverty rate estimates, labor statistics, FBI crime statistics, and building permits.

The **Fair Market Rents** dataset shows the fair market rents, which HUD estimates annually for each county in the United States. The **HUD Median Family Income Limits** dataset contains estimates of income limits for different family sizes at the county level. The income limits are used to determine the income eligibility of applicants for public housing, Section 8, and other HUD programs. The **Annual Adjustment Factors** data for each metropolitan area are determined by a formula utilizing information such as consumer price index and residential rent and utilities cost changes. This information is used to adjust contract rents for units participating in HUD's housing assistance programs.

The **Assisted Housing** dataset sketches a picture of nearly 5 million subsidized households across the United States. Included are housing variables such as the total number of subsidized households, as well as demographic variables such as household income and number of children. The data present U.S. totals, state totals, and census tract summaries. Data are also summarized by local public housing agencies and by individual housing project.

The **Government-Sponsored Enterprises** (Fannie Mae and Freddie Mac) data contain information on mortgage purchases of Fannie Mae and Freddie Mac. The data are tabulated at the national, state, MSA, and census tract levels. The data will be useful to studies of the flow of mortgage credit and capital in American communities where Fannie Mae and Freddie Mac are focusing their affordable homeownership efforts.

Finally, the **Research Maps (R-MAPS)**, volumes 2 and 3, datasets contain a portion of data selected from the above list and made into a GIS-readable format (such as ArcView and LandView). These two volumes represent an effort by the Office of PD&R to make the above data available in spatially enabled data format (see Box 2.4).

Data Needs and Issues

PD&R has taken significant steps in collecting data and making data available to the public. HUD's datasets are widely used, by researchers from universities and policy agencies to town governments and community advocacy

---

BOX 2.4
R-MAPS

PD&R designed Research Maps (R-MAPS) to make HUD data more accessible and useful to researchers, policy makers, and practitioners. The data were initially provided free on CD-ROM but are now provided on-line as part of HUD's EGIS. The datasets are presented as shape files that link tabular data to boundary files. Spatial query and analysis tools are provided through LandView. Data available through this software include American Housing survey data, government-sponsored enterprise and Home Mortgage Disclosure Act information, Low-Income Housing Tax Credit locational data, housing data, data on public housing and project-based program areas, and data reported in the State of Cities publications.

---

groups; however, a number of data issues should be addressed to make the disseminated data more useful. The issues of data quality and completeness are paramount. Data from different sources often need to be integrated and "cleaned" spatially and thematically[11] to maximize their utility.

The data development process is the most costly component of a GIS, accounting for about 60 to 85 percent of the cost for most organizations (Longley et al., 2001). The success of an enterprise GIS (see Box 2.5), and therefore its organization, is determined by its ability to provide high-quality and useful data. Data quality also determines the accuracy of subsequent GIS analyses. Error in data tends to propagate through the analysis steps, making subsequent analysis results unreliable. For example, to produce aggregate measures of the incidence of poverty in small areas within cities, HUD uses census data on household income and household composition. Data on rental housing cost may then be used to produce measures of the proportion of household income spent on rent. Error in any of these datasets will be propagated along the chain. Although GIS is useful for storing, retrieving, and analyzing data, it cannot assure data quality, which is essential for sound policy and planning.

In evaluating the usefulness and quality of a spatial dataset, a number of factors should be considered. These include: the geographic scale or extent (national or local coverage), the completeness of the coverage (gaps or holes),

---

[11] Spatially, when different datasets are combined, boundaries and roads may be topologically inconsistent and require matching. Thematically, two datasets may have different attributes or coding and necessitate matching or filling in of missing attributes.

the spatial resolution (level of spatial details), the attribute accuracy (e.g., sampling density), the frequency of updates, the degree of confidentiality, and finally the compatibility and comparability with other data sets.

Another consideration is the perpetual tradeoff between data quality and its development cost (time and personnel needed). Cost is generally higher with higher data quality. Unfortunately, data error is impossible to entirely eliminate, and it will always exist. Therefore, it is important for an organization to maintain a balance between cost and quality. The objective in this case is to "manage" rather than try to eliminate the error by acknowledging some degree of uncertainty through various statistical means (Aronoff, 1989).

The above synopsis of the 13 datasets shows that there are rich sources of data that can be utilized for housing research; however, it reveals three issues. First, although most of the existing datasets have some geographical identifiers, not all have been fully spatially enabled. These datasets will have to be address-matched so that the user can retrieve and map information by its spatial location (e.g., a city, a tract). Second, the smallest geographical areas reported for most datasets are at the city or county levels; a re-tabulation of the data into smaller geographical areas, such as at the census-tract scale, should be considered to enable micro-level analysis. Smaller geographical areas such as census blocks or block groups will be more desirable, but they are also more vulnerable because of confidentiality breeches. However, census tracts are an appropriate geographic scale as a first step for a nationwide coverage and some of the existing data are already reported at this geographical scale. Third, some data are incomplete, and these gaps must be identified and filled.

Conclusion: HUD datasets derive from a multiplicity of sources. Local datasets can be a valuable source of accurate and detailed data that is relevant to HUD's local constituents.

Recommendation: As a first step, HUD should improve existing housing and related data. Existing data should be cleaned and checked for accuracy, consistency, and completeness. Data gaps should be identified and filled. HUD should adopt accuracy and documentation standards that build on FGDC data standards.

Building on existing data is only the first step. As GIS capability improves, new national datasets as well as local, finer resolution data will be wanted, so that new possibilities for research can be realized, and a response to societal needs can be accommodated in a timely and accurate manner. New data development can be concomitant with research issues that are high on HUD's research agenda, such as home ownership issues, the housing conditions in the Colonias on the southwest border, and growth management problems for communities (see Box 3.1). HUD's current GIS initiative, entitled EGIS, is described in Box 2.5. An internal spatial data infrastructure or an agency-wide GIS can provide a common platform that facilitates data use and dissemination. An internal GIS infrastructure is an enterprise GIS implemented as an organization-wide platform to structure the collection, storage, analysis, and presentation of geographic information. To accomplish this, a governance structure must be created to adopt standards, implement consistent business practices, and develop an organization-wide strategy for this purpose.

BOX 2.5
Enterprise GIS

In 2000, HUD entered into a contract with Environmental Systems Research Institute (ESRI) to build an Enterprise GIS or EGIS. The EGIS allows agencies of all kinds and at all levels, and the public in general, to view HUD housing and community development data together with data from three other federal agencies (U.S. Census Bureau, Environmental Protection Agency, and the Federal Emergency Management Agency). The Census data include both 1990 and 2000 data, providing a temporal element to HUD's GIS platforms.

HUD's addition of metropolitan-level data from other agencies represents an expansion of capability beyond that of earlier systems. By accessing the EGIS web site, individuals and community groups can combine data sets in different ways to compile a rich base of information that is specific to the user's needs. The information available through the EGIS includes spatially-referenced data on multifamily housing, brownfields tax incentive zones, public housing, hazardous waste and air pollution. Using the EGIS, users can:

- Create a personalized map. Users may enter an address or click on a map and have the application take them to a map of the location.
- Add data to maps. Users may display any combination of HUD Housing and Community Development data, along with data from any of HUD's federal data partners (EPA, FEMA, Census), as well as data layers that include lakes, rivers, landmarks, city streets, highways, and other features
- Save maps. Users may save the maps they create, name the map, then retrieve it whenever they use EGIS again
- Create a thematic map. Users may create a thematic map based on data and criteria that the user specifies, thus allowing the user to

---

classify data into groups or classes that have similar characteristics and values; tables associated with the maps will also be available
- Print maps. Users can print the maps they create, and use them for their own purposes.

The EGIS is in its early stages, but several challenges have been identified (Mark Mitchell, ESRI, personal communication, 2002). First, there is a need for data integration. Data compatibility is an important issue because of the many data sets that comprise the EGIS, and their varied sources. Second, there is a need to improve data quality generally. Third, early experience with EGIS points to a compelling need to track funds allocated by HUD and create a more rigorous tracking and assessment program to determine the impacts of various interventions. Fourth, the public still has difficulties gaining access to good information. Knowledge of GIS and spatial analysis among local governments and community groups—especially in smaller cities and towns—is often lacking.

See <http://www.hud.esri.com/egis>.

This infrastructure must have authorization and support from the Secretary. It should be created and managed within PD&R, but the changes in business practices required will affect all data gathering, storage, analysis, and presentation within the department and local housing agencies. It requires a long-term commitment of financial and human resources within the department and local housing agencies, but will permit, *inter alia* geographic analysis of the following:

- Strength of prior HUD investments;
- Effect of HUD investment on the stability of neighborhoods, municipalities, schools, and school districts;
- Educational and economic opportunity present in areas of potential HUD investments; and
- Future investment decisions that will foster health, education, and economic opportunity, and residential and commercial stability of neighborhoods and regions.

Conclusion: A spatial data infrastructure can provide a uniform and high quality of service delivery across HUD's programs and missions. GIS can foster agency-wide data coordination, integration, sharing, and analysis; and facilitate internal assessment of HUD programs, and analysis and reporting of federal urban investments. An integrated spatial data infra-structure can aid in the delivery of services to HUD clients such as metropolitan or regional organizations and local governments. It can also enable local and regional information to be integrated in ways that allow for more accurate program assessment and for evaluation of federal investment in urban development.

Recommendation: HUD should create an internal spatial data infra-structure for an agency-wide GIS to support an appropriate urban research agenda and to integrate locally derived data.

Integrating Local Datasets

HUD's 81 field offices nationwide represent a rich source of local data and a wealth of relationships at the local level. Local governments submit their data to the federal government and receive data back in the form of TIGER/line files[12] and other products including a digital database of geographic features. The local data that go into the creation of these products can surpass the resulting TIGER/line files in terms of local accuracy and relevance. When local data users find discrepancies in the returned TIGER files (for example, in the spatial boundaries of their local areas), they may spend significant time updating and modifying the TIGER files to accurately represent known local conditions. No mechanism is in place to integrate these updated and accurate datasets from local users and to re-distribute those datasets to other users or to integrate them into a national database.

For most urban areas, TIGER data are derived from dual independent map encoding (DIME) files, originally developed for the 1970 census. DIME is an encoding scheme for street addresses. Integration of local (neighborhood and parcel-level) data with federal data from HUD and other agencies will be facilitated by new systems and technologies. At present, however, such data integration represents a formidable challenge. Similarly, efforts to improve data quality constitute a major investment whose full range of costs and benefits are not known. The incorporation of comparable local data would make data available at multiple scales on a broad range of urban topics including real estate market conditions, neighborhood educational and economic opportunity, crime, the quality of local housing stock, and environmental risk. These data could be disseminated via the Internet, saving HUD data users the

---

[12]Topologically Integrated Geographic Encoding and Referencing System (TIGER). See <http://www.census.gov/geo/www/tiger/index.html>.

time-intensive work of data integration. The long-term commitment needed for such an effort would produce data for national comparative analysis at a resolution useful to local agencies.

As the prime unit at HUD responsible for providing reliable and relevant data for research and analysis, PD&R has an important role to play in promoting data integration and data sharing within the agency, and between HUD and its partners. Consistent data from all internal HUD units and HUD partners are needed for coordination and optimal use of data.

Conclusion: As a result of the agency's local relationships, HUD has significant access to local data and a singular ability to mandate national standards for local data. The HUD grantee program is another valuable source of local data.

Recommendation: HUD should develop mechanisms to accept and integrate relevant locally derived data and georeference the data for integration in the agency-wide GIS. Specifically,

- **HUD should spatially-enable local data by performing address matching of individual records at the finest scale using geographic coordinates.**
- **HUD should select, tabulate, analyze, and map relevant housing variables through a GIS at multiple relevant geographic scales (census block, block group, and tract; place, county, and metropolitan area).**
- **PD& R should take the lead within HUD in efforts to integrate grantee and other data at different levels: parcel, neighborhood, municipality, school and school district, metropolitan area, state, and national.**

THE NEED FOR AN URBAN SPATIAL DATA INFRASTRUCTURE

HUD's challenge—using spatial data to promote adequate and affordable housing, economic opportunity, and a suitable living environment free from discrimination—is considerable in scope. In its mission, HUD identifies six related strategic goals (HUD, 2002b):

- Increase homeownership opportunities,
- Promote decent affordable housing,
- Strengthen communities,

- Ensure equal opportunity in housing,
- Embrace high standards of ethics, management, and accountability,
- Promote participation of faith-based and community organizations.

These goals emphasize the need for HUD to act responsibly and to encourage community participation in the urban and housing arenas. The recommendations outlined in this chapter include full participation in the FGDC and other federal initiatives; ensuring the accuracy, consistency, and completeness of HUD data; creating an internal spatial data infrastructure; and developing ways to integrate and disseminate local data. These goals are encapsulated below in the concept of an Urban Spatial Data Infrastructure (USDI) as a component to the NSDI for urban areas.

Because of its relationship with groups at the local level across the nation, HUD has unique access to detailed, updated data on local conditions and local needs. Local data are needed by HUD to address its agency mission and, when stored in a national database like the NSDI, are useful for multiple purposes including comparative urban analysis, resource and services allocation, and homeland security. At this time, the United States does not have a source of standardized local, parcel-level data. Figure 2.2 shows the correspondence between HUD's agency mandates and national data needs.

The Bureau of land Management (BLM),[13] through its cadastral survey, is responsible for the identifying the location and boundaries of federal lands in the United States. The agency maintains cadastral survey and historical data, along with information on the mineral estate, resource conditions, and permits or leases on federal lands. BLM's Geographic Coordinate Data Base (GCDB) is using GIS to modernize data management for a parcel-based land information system that meets FGDC standards. Initially data were collected in the western states where most federal lands are located; collection is now proceeding in the eastern states. BLM also has a successful online data distribution system of the Public Land Survey System, but these data are also limited to the western states.

The creation of a nationwide parcel-level dataset will require the participation of local government, finance agencies including Fannie Mae and Freddy Mac, realtors, and market researchers. States and metropolitan/ regional-level governments (for example, the Twin Cities in Minnesota) have created programs to create or modernize parcel-level data. Because there is no nationwide source of parcel-level data, costly duplication and gaps can occur. In its effort to provide accurate and relevant data on urban and housing conditions, help the homeless, spur economic growth in distressed neighbor-

---

[13] See <http://www.blm.gov/cadastral/> for details.

hoods, and help local communities meet their development needs, HUD is undertaking an effort equivalent in value and significance to the creation of an urban spatial data infrastructure (USDI) for the nation. For HUD, as for other federal agencies with responsibility for providing spatial data for national initiatives, carrying out this effort demands significant time and resources.

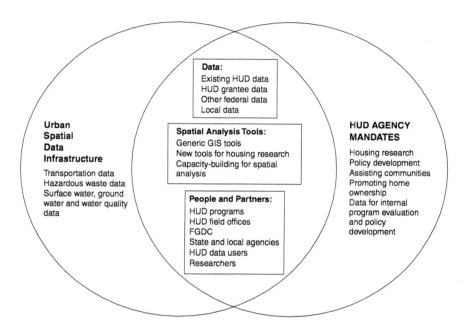

FIGURE 2.2   This Venn diagram demonstrates the overlap between the building of a USDI and HUD's agency mandates. The data, tools, and partners needed to carry out HUD's mission and support agency research are the same as those needed to build a USDI.

HUD shares responsibility for the provision of these important urban data with other federal agencies, notably the Department of Health and Human Services, the Department of Transportation, and the Environmental Protection Agency.   The task of cross-referencing data at this fine-grained scale is complex and spatial features used by various agencies do not correspond one-to-one.   For example, parcels and boundaries do not mesh with buildings, blocks, streets and street centerlines are not the same as bus-stop-to-bus-stop routes. Floodplains and atmospheric plumes find their own paths. Data sharing in such circumstances involves standardizing data formats, semantics, and syntax. Standardizing 911 emergency addresses and developing master address files is an example of an ongoing multi-decade effort to facilitate address-based geo-referencing. The importance of partnerships and communication among the federal agencies is discussed in Chapter 5.

For many purposes, census tracts are too aggregated or too sensitive to problems associated with arbitrary geographic units, known as modifiable areal unit problems. On the other hand, parcel and housing unit data are often too disaggregated, cumbersome, or invasive of privacy for many analyses. HUD has a potential role to play in the development of relevant intermediate layers. In particular, HUD can influence the development of standardized procedures for computing land value surfaces, housing price indices, job accessibility measures, and other derived data layers that are finer grained than census tracts but appropriately aggregated and smoothed compared with parcel maps. HUD could promote data cleaning, interpretation, and statistical analyses needed to develop some of these specialized intermediate layers and play a lead role in making these data a meaningful and reliable component of urban models. The use of software tools to aggregate and adjust the individual data into a customizable form could facilitate the use of confidential data that may be more useful than data at the census tract level. Eventually, such estimated intermediate layers could be derived "on-the-fly" by HUD-provided online tools.

Conclusion: HUD is well-suited to be one of the lead federal agencies in providing and managing urban framework data layers for the NSDI. HUD functions within a network of agencies at multiple levels that share responsibility for providing data on urban and community issues.

Recommendation: HUD should promote the development of a parcel-level data layer and other urban framework layers to create a USDI as a component of the NSDI for housing and urban development. The federal government should make available resources commensurate with this task.

Core elements of the USDI can include:

- Public and federally assisted housing data,
- Tenant and housing characteristics,
- Parcel-level data,
- Locally updated TIGER files,

- Environmental data, and
- Socioeconomic data.

Examples of socioeconomic and environmental data include hazardous waste and "brownfield" site location data, crime statistics, transportation data, health data, educational data, and business activity data. Health data include mortality, morbidity, and immunization statistics. Education data include school performance, enrollment, and percent receiving subsidized lunch. Business activity data include the number of business establishments, and employment data.

The development of GIS at HUD is not so much a question of purchasing GIS software systems and data as it is the development of appropriate, maintainable databases, information infrastructure, and interagency-agency relationships. HUD must upgrade its internal spatial data infrastructure and play a more active role in shaping an urban spatial data infrastructure (USDI). Rather than a system that HUD owns or manages, the USDI would include many agencies and public-private partnerships at multiple levels of government. The USDI would comprise locally managed metropolitan information infrastructures that can feed appropriate data to HUD and use HUD-processed data, analyses, indicators, and models to improve the understanding of local conditions and the design, delivery, and evaluation of local plans and services.

Conclusion: The collection and dissemination of relevant and accurate urban data require close coordination with state, regional, and local groups. Communication with local groups ensures that the data collected are meaningful to the community. Partnerships with local groups build capacity for research and applications and promote the collection of accurate data that meet federal standards for data sharing. Local data centers can support the development of a USDI.

Recommendation: HUD should encourage and support the development of local, metropolitan, and regional data centers to facilitate local data coordination, use, and training towards the creation of a USDI.

SUMMARY

Developing, maintaining, and disseminating reliable spatial data are major challenges for HUD. HUD's efforts will be more efficient and productive when carried out in coordination with other federal data initiatives, notably the NSDI. To provide accurate and consistent data on housing and urban issues and use these data for internal evaluations of HUD programs and investments, HUD needs an in-house spatial data infrastructure. The challenge that HUD confronts in use of spatial data to address its broad and important mission goals demands an effort that is tantamount to the creation of an urban spatial data infrastructure (USDI) as a component of the NSDI. The integration of local datasets into HUD's databases can provide relevant, high spatial and temporal resolution data for HUD's internal program analysis and evaluation, and for national data and information needs including the creation of a USDI, resource management and allocation, and other federal data initiatives. These are formidable challenges that resist simple solutions. This chapter offers a vision of the future of GIS at HUD that demands administrative support and adequate resources for spatial data initiatives.

3

Data Dissemination and Software Tools

To provide relevant and accurate data on housing and urban issues requires balanced attention to a number of related efforts that will encourage access and use of the data. These issues include promoting broad public access to data and information by building local capacity to use data, and supporting the development of analytical tools and skills (in-house and among the agency's clients) that enable spatial analytical research on the complexities of neighborhood and urban issues.

Privacy considerations are also critical for data dissemination and application. Chapter 2 discusses the need to incorporate local data into national databases and to get that data back to users for comparative analysis. Unless methods are developed to disseminate data at required resolution while maintaining privacy, local groups will be unwilling to share their data. This chapter describes the wide array of HUD data users and discusses methods for gathering and disseminating these data, and for developing related tools to support decision making and spatial analysis.

HUD DATA USERS

HUD data are important to a broad spectrum of users. Users include: the agency's staff at national and regional offices and public housing authorities; policymakers at federal, state and local levels; community-based and advocacy organizations; private researchers; university faculty and students;

tenant associations; and community residents and businesses. The data are used to plan community development and services, prepare grant applications, conduct research, implement programs, and support advocacy efforts.

HUD serves low-income and minority populations in the nation and it is important to include these groups in decision making about their community. How can HUD get the tools of GIS into the hands of citizens so they get involved in their local planning processes? How can HUD link available analytical and decision-making tools with local knowledge to address local needs and priorities in housing and urban development? How can HUD promote spatial analysis of urban and housing issues in its client communities and among its in-house staff? These are among the agency's challenges in which PD&R plays a central role.

GIS can be used in HUD's 81 field offices throughout the United States in a number of ways: internally by HUD for many purposes including linking its far-flung field offices, by recipients of HUD grants, and by HUD-related advocacy groups. At present, however, the use of GIS by HUD grant recipients (for instance, entitlement communities) is limited despite efforts to make the technology available (e.g., Community 20/20). Often, these groups are disadvantaged by being disconnected from both the planning and the information technology divisions of local government (Michael Martin, U.S. HUD Milwaukee Field Office, personal communication, 2002).

HUD grantees have tended to use GIS to identify the spatial distribution of their programs, to create visual displays of resource allocation for political and educational purposes, and to advocate programmatic directions. Internally HUD's regional offices use GIS in a variety of ways, including: ascertaining the eligibility of localities for place-based HUD funding, documenting HUD's investments, investigating fair housing, and responding to disasters. Most of the GIS efforts in HUD's field offices do not go beyond point and thematic mapping, because of limited understanding of spatial analysis, comparative spatial statistics, and housing indicator development (Michael Martin, U.S. HUD Milwaukee field office, personal communication, 2002).

Typically, the use of GIS in HUD-related advocacy has revolved around education and communication with traditional HUD intra-agency groups created to help build support for HUD projects and activities. In particular, HUD has used GIS with organizations that promote fair and affordable housing. Specific projects include planning and disseminating information. This includes: answers to the question, "What does HUD fund in my community?"; building a community consensus on affordable housing needs; promoting better understanding of local real estate investment; and multi-media GIS to visualize, and to help non-experts visualize planning project alternatives. Used in these ways, GIS can be an effective tool for encouraging community engagement in decision making and planning.

Visualization, Communication, and Power

"If information is power ... and if community is built through dialogue, then informatics permits both to emerge for those who would otherwise have no voice and no space for collective action" (Pickles, 1995, 10).

Local organizations can become powerful players using GIS. Although local people may not think about their neighborhood in terms of GIS data layers, they may consider community issues comprehensively, in an integrated way in which "the quality of life in [the] family and [the] neighborhood revolves around housing, and work, and safety, and education, and goods and services, and neighbors and social networks and... many other things" (Michael Barndt, Nonprofit Center, Milwaukee, personal communication, 2001). Crime, transportation, housing, health care, families, and many other issues are in the minds of community actors as they think about their neighborhoods and attempt to develop solutions to meet neighborhood needs. GIS is a tool that can help integrate and analyze complex information and display it with the visual clarity of a map.

Currently, at the local level, GIS is used largely for mapping or visualization to identify problems, show them spatially, and use the maps to advocate for public policy changes. Mapping serves several functions. First, a map can transmit new information to community residents in a way that is visual and easy to understand. For example, seeing and showing that crime is concentrated in one area of a neighborhood sends a powerful message. Residents can take a map to city hall or the city council to argue for additional resources for their communities.

Community-level affordable housing advocate, Stella Adams, stresses the importance of spatial data to empower people to advocate for themselves (Stella Adams, North Carolina Affordable Housing Center, personal communication, 2002). Using HUD housing data, Resident Advisory Councils can go online and see what is going on in their community. Ideally, these HUD datasets would incorporate local knowledge from these groups within the framework of a larger dataset. Adams says, "Having access to GIS empowered me to be able to do things I couldn't. [People] need the proof to show to elected officials. A map gives them legitimacy. Then you're paid attention to."

Many of the decisions regarding housing and urban development are ultimately made by local municipalities and by states although HUD provides resources and develops the national housing and urban agenda. HUD has long sought a strategy that would support local efforts and provide tools for communities, local governments, and other interested partners to use GIS and spatial data for local planning. Community 20/20 software was designed to enhance access to geographic information, to put HUD in the forefront of using spatially enabled technology, and to facilitate local planning processes and community efforts to prepare their consolidated plans (Box 2.2). HUD's follow-up program, Enterprise GIS, continues this effort (Box 2.5).

Understanding Local Conditions and Needs

Informal settlements along the U.S.–Mexico border (colonias) are a good example of the need for understanding local conditions, for attention to processes at work at the regional level, and for putting relevant geographic information in the hands of local decision makers. HUD could use GIS in the colonias to evaluate the question the appropriate scale for the analysis of urban problems, assess the regional labor markers, identify housing and rental prices as a determinant of colonias development, monitor transnational processes between the United States and Mexico, and analyze the changing economic practices of unregulated urban settlements (see Box 3.1). Aerial photographs make effective visual tools because customers can "see" the area in question and better understand the other data that is overlaid on the map. The partnership between HUD and the U.S. Geological Survey in the colonias has made good use of these photos (Figure 3.2).

Although visualization is a powerful tool for community groups and policy makers alike, the complexity of some urban problems requires advanced analytical techniques including statistics and modeling. Spatial analysis can inform our understanding of social problems and suggest public policy response. For example, understanding private housing markets is critical to housing voucher use. Housing vouchers work well in urban areas where there is a surplus of housing and landlords are happy to have the opportunity to rent their apartments through HUD's voucher program. In tight housing markets, landlords have little incentive to accept housing vouchers. When demand for housing exceeds supply, prospective tenants bid up the price of rental units. In these situations, landlords often prefer to let the market work, since rising demand leads to higher rental prices. Individuals who have vouchers in these situations may find that they cannot secure a place to live. GIS could provide a framework for keeping track of trends in the private housing market including rental prices, and lead to the development of additional means to provide affordable housing in urban areas.

GIS can be used to analyze the availability and spatial distribution of housing for people with low income. Nationally, the availability of affordable housing is lower in the suburbs than in the cities. In the 1990s, housing costs rose faster than family income (Lichter and Crowley, 2002). Although housing economists at HUD currently track these trends at the national level, tracking at the city and neighborhood levels will provide additional information. Availability and dissemination of local data are key issues in integrating the fine-grained detail of housing market issues within the larger context of national trends. Analysis that integrates data at the local, regional, and national scales is required for the development of urban public policy. Confounding the integration of analysis at various scales is the lack of digital data and GIS expertise at the local and neighborhood levels. The role that HUD can play in promoting the inclusion of local data in national databases is discussed in Chapter 5.

A Spectrum of User Needs

Users of HUD data vary in terms of technical ability and access to resources. Regardless, most people want timely, accurate, and accessible information. People want information about their neighborhood, such as the availability of homes to rent or buy, and the location of social services and transportation routes. More advanced HUD data users want this information in the form of spatially enabled data with accessible metadata to show, find, and explain interesting trends and patterns.

BOX 3.1
Colonias: U.S.–Mexico Community-based GIS
for Economic Development

HUD defines the colonias as "rural communities and neighborhoods located within 150 miles of the U.S.–Mexican border that lack adequate infrastructure and frequently also lack other basic services.[1] Colonias have emerged in rural areas but they are predominantly residential areas for workers and families working in nearby urban centers or in agricultural occupations. Colonias vary in age, size, and composition, but because of their informal nature and recent origin, little is known about the trajectory of their growth. Typically they are unplanned, unregulated settlements with high rates of poverty, which is a factor that compounds the difficulty of developing infrastructure such as roads, water and sewer systems, improved housing, street lighting, and other services.

---

[1]Definitions of a colonia vary among agencies and groups (<www.hud.gov/whatcol.cfm>).

BOX 3.1 Continued

In the 1990s, Congress required states along the Mexican border to set aside a portion of their HUD-allocated Community Development Block Grant (CDBG) funds to alleviate poverty and improve housing in the colonias. Other federal agencies, including the EPA and Department of Agriculture, also have projects that deal with the colonias.

FIGURE 3.1 House in the colonias. SOURCE: Alina Simone, Texas Low-Income Housing Information Service.

HUD's goal in its colonias project is twofold:

- understand the conditions within the settlements to identify emerging issues and challenges, and to inform PD&R and HUD's responsibilities in the colonias, and
- to carry out this research without doing extensive on-the-ground fieldwork as a demonstration of the potential usefulness of GIS approaches to urban problems.

Monitoring the colonias along the U.S.–Mexico border is being undertaken in conjunction with the U.S. Geological Survey (USGS), through an interagency agreement. This agreement seeks to develop a joint HUD–USGS cooperative GIS-enabled web site. The demonstration site for the project is the city of Eagle Pass, Texas. Three other pilot communities, one in each of the remaining Mexican border states, will be developed based on the experience gained in Eagle Pass.

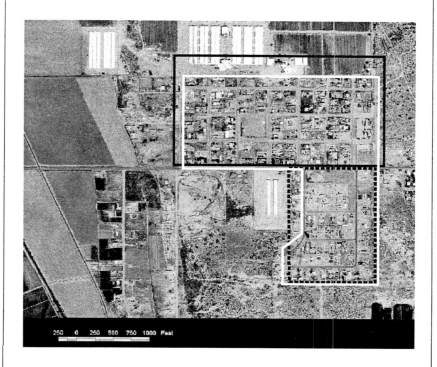

FIGURE 3.2   Aerial photograph of colonias in Berino, New Mexico. Note the multiple boundary lines pictured (black line, dotted line, white line) because of the lack of a consistent geographic definition of a colonia.   SOURCE: Robert Czerniak, Department of Geography, New Mexico State University.

PD&R is in the process of overlaying USGS aerial photos and data on housing and water facilities and integrating these layers into HUD's enterprise GIS platform. Data from other agencies, such as the Department of Health and Human Services, EPA, and the Census Bureau will subsequently be integrated.

Technically sophisticated users may want substantial flexibility in querying and downloading HUD data; access to confidential data; and the ability to integrate multiple data sources, integrate data from different time periods, and use the data for multivariate spatial analysis. Less technically sophisticated users and users with few resources may have different demands such as easy-to-use tools to access HUD data for making maps or charts and tables that address their needs. Such users will need an Internet-delivered or web-based interface so that they can access and use HUD data at their local library or community center using basic hardware and software. The level of expertise and the needs of users are important considerations for the design of web-based GIS. HUD's EGIS is setting out to address these user needs.

Conclusion: HUD is faced with a broad a spectrum of user needs from the basic to the more technologically advanced. The full range of users must be considered. A web-based interface is important for some of HUD's clients and for certain applications but the quality and usability of data are essential for all clients and all applications. In addition to simple, web-based interfaces, flexible querying is required to support more sophisticated applications.

Recommendation: HUD should continue to develop a spectrum of tools to meet user needs.

- **For users with limited financial or technical resources, HUD should provide web-based mapping of HUD data and related information.**
- **For more advanced applications, HUD should develop tools for flexible querying, extracting and downloading data, including standard file formats for exchanging data.**

DATA DISSEMINATION

HUD disseminates data through multiple avenues including the Internet, HUD USER, and academically affiliated and unaffiliated researchers. Datasets and dissemination strategies are not centralized. Information disseminated to HUD's local and regional offices is not always adequate or easy to use. HUD's current methods for disseminating data are discussed below.

HUD's Enterprise GIS (EGIS)

The EGIS is a portal to access HUD's data and to coordinate relevant data from other private and public agencies either through direct access or links. In its current state, it provides limited use to more advanced researchers, but the EGIS is useful for visualization purposes. Users can map locations and add layers for political and program boundaries, and add point data from HUD or other programs. More sophisticated users can download data to their own personal computers but EGIS does not support analysis of urban and housing issues at different scales or geographies (see Box 2.5). Using GIS effectively is a complex goal involving HUD's information and communication technology (ICT) infrastructure, personnel skill and training, organizational structure, partnerships, and technology choices.

HUD's R-MAPS

R-MAPS, PD&R's packaging of 13 spatially enabled datasets provides a wealth of HUD-related housing data. The Guide to PD&R Data Sets[1] includes basic information about each of the datasets, but to get a full list of variables, variable definitions, and other specifics, users must download data directories and other supporting documents, which makes the use of the datasets cumbersome for sophisticated users and impossible for less sophisticated users (Box 2.4).

Custom Data Tools

To disseminate data, HUD has invested in customized tools like Community 20/20. Unfortunately, customized tools are limited in capability compared with the commercial and public domain or open-source tools. HUD would do well to avoid further custom tool development for data dissemination. Online data tools that support a broad range of users will promote the use of geographic data.

Research Clearinghouse

HUD can build relationships with research communities to facilitate use of spatially enabled data to examine housing and urban issues. The agency

---

[1]Available at <http://www.hud.gov>.

has taken steps to initiate similar information sharing resources, for example, the Regulatory Barriers Clearinghouse.[2] Along with the creation of a USDI, described in Chapter 2, HUD can support the development of a node for urban and housing research in the National Geospatial Data Clearinghouse of the NSDI. A good model is the USGS node for information about geospatial or spatially referenced data.[3]

Strategies for Developing a Research Clearinghouse

Discussed below are several strategies to develop a research clearing-house for housing and urban issues. These include better documentation, support for user conferences and online data user groups, a HUD USER newsletter focused on spatial analysis, and training to develop spatial analytical capabilities both in-house and for local data users. HUD could use the research clearinghouse to encourage researchers to examine urban and housing issues at different geographic scales and work closely with communities to develop research questions and create research products that are useable by communities, non-profits, and other local groups.

As HUD makes additional housing-related data available via the Internet, the agency could further facilitate researchers' access to and use of these datasets by creating better documentation that includes accurate and complete metadata. HUD could also support conferences on urban and housing themes or support particular tracts of panels within existing conferences to promote the understanding of the importance of data standards and to encourage the advancement of spatial methods to examine urban questions and problems. The Internet permits the formation of online data user groups. The American Housing Survey offers an example.[4] HUD could encourage the formation of online user groups around the use of HUD data sets and discussion of broader urban and housing issues.

The HUD USER newsletter is another means of disseminating information about current urban and housing research and the availability of relevant data and tools. HUD could create a new newsletter devoted to spatially enabled urban and housing research, methodologies, and issues about datasets. A newsletter could provide a source of information on HUD funded projects and other related projects including descriptions, implementing agencies, datasets employed, and status reports.

---

[2]See <http://www.huduser.org/regbarriers/>.
[3]See <http://nsdi.usgs.gov>.
[4]See <http://www.huduser.org/datasets/ahs.html>.

Finally, HUD's public includes highly sophisticated users with significant resources and users with far fewer resources. Additional training is necessary to bring resources to less sophisticated and resource-poor users. Many of the institutions, groups, and individuals who should be part of discussions about housing and urban issues lack access to technology and training to fully utilize GIS. Community organizations, neighborhood residents, and even HUD field office staff require both technical training and products that are easy to access and use.

HUD's Community Outreach Partnership Center (COPC) Program

Many of the tasks involved in disseminating and using data to analyze urban conditions are best done locally. Regional or metropolitan centers can work collaboratively with communities to inform research questions and agendas and to collect and present data in ways that are useful to the community. These centers and others, such as State Census Data Centers, can serve as regional or metropolitan points to gather data across time periods, negotiate partnerships with other data producers, clean data, and disseminate it. A number of Community Outreach Partnership Centers (COPCs) have begun this work.

Taking advantage of Environmental Systems Research Institute's (ESRI's) Arc Internet Map Server (ArcIMS), a product that allows for the development of interactive-mapping web sites, a number of COPCs have developed web sites for the community and city in which they work. Communities produce maps to meet neighborhood needs. Researchers post commonly used data such as U.S. Census data along with a variety of boundaries such as political jurisdictions or census geography, and users can create their own maps. Neighborhood Knowledge Los Angeles (NKLA), developed by graduate students at University of California at Los Angeles's COPC,[5] is one of the more sophisticated sites. The site allows users to post their own data and projects, and use data posted by the university. Users can also post the results and descriptions of the projects for which they have used data, in the process building a library of examples to generate future ideas.

These COPC web sites are what are known as "thin" GIS sites. They provide data for mapping or visualization directly on the site. They do not necessarily encourage more sophisticated analysis that would include analysis of problems at multiple scales or downloading data for multivariate analysis. HUD's COPC program is structured to emphasize outreach rather than

---

[5]<http://copc.sppsr.ucla.edu/>.

research. Intermediaries would be needed to develop research capacity in the community. The potential for cooperative centers can be seen in the example of the Nonprofit Center of Milwaukee, a coalition with more than 200 local members (including the local HUD field office) that created the Community Information System (Box 3.2).

Most of the COPC projects rely on interactive mapping to distribute data. Providing access to data over the Internet reduces the level of technical sophistication necessary to create basic maps; however, users who lack Internet access or have slow Internet connections are at a disadvantage. Furthermore, users are limited to available data. Finally, map-making is part art, part science, and a lack of expertise can result in maps that distort the data they are intended to present.

BOX 3.2
Milwaukee's Community Information System

The Community Information System in Milwaukee, Wisconsin, facilitates access to data and builds the capacity for community and non-profit organizations' data use. The center provides access to data and training in the use of data; and works closely with its partners to build a sustainable neighborhood data clearinghouse, offer data and GIS services on demand to neighborhood organizations, build the capacity of local organizations to organize and interpret data, and use technology to create tools to lower the costs of accessing and analyzing data. Rather that log on to a web site, community users work closely with the expert staff of the center to identify research questions, negotiate partnerships to share data, develop maps to present the data, and build capacity to use the data and maps to influence public policy and address urban problems and issues.

The Role of PD&R in Data Dissemination

As outlined earlier in this chapter, PD&R plays an active role in the interagency work that HUD is doing in the colonias, in relationships with other HUD clients and partners (such as urban researchers and community groups), and in dissemination of spatially enabled and housing-related data with R-MAPS. In the past, some GIS efforts at HUD such as Community 20/20 were hampered by a lack of technical input. Updating and maintaining data for Community 20/20 may have proved difficult in part because no program office at HUD had clear responsibility or ownership of the initiative.

PD&R's current active role in providing technical expertise, research, and analysis for the development of HUD's enterprise GIS is necessary and appropriate.

Conclusion: A user friendly, web-based GIS is an efficient means for providing information about local housing conditions and making basic data available to the public. The utility of the information depends on the accuracy and the relevance of the basic data and the methods by which the information was derived. The development of a well-designed web-based GIS is a long-term process. User input is critical to this process.

Recommendation: To improve dissemination and promote the use of spatial data, HUD should:

- **Involve users in design of the web-based GIS;**
- **Sponsor conferences and workshops for clients and partners about using spatial data;**
- **Support online groups for HUD spatial database users; and**
- **Produce an Internet newsletter devoted to spatial data and analysis.**

PD&R is well-positioned to:

- Work with HUD clients and data users to derive the most appropriate GIS designs and to identify needed data and functions.
- Manage data confidentiality. For some sensitive data, PD&R will need to develop a policy on releasing confidential data as well as algorithms to suppress sensitive data to protect privacy.
- Take a lead in establishing a node for housing and related economic and demographic data in the NSDI's National Geospatial Data Clearinghouse.
- Support the functions of an agency-wide enterprise GIS across all relevant HUD units.

HUD GIS SUPPORT TOOLS

GIS Applications and Needed Technological Support

HUD's mission has a strong spatial component so there is tremendous potential and need for spatially referenced data and geographic analysis for policy development. For example, one can use GIS to determine mortgage distribution, patterns of segregation, and neighborhood change. GIS has multiple capabilities and each is valuable in its own way, for basic mapping, data handling, spatial analysis, etc. Local governments and community groups are beginning to take advantage of GIS and are developing capacity to do more sophisticated spatial analysis.

GIS technology encompasses three integrated components at different levels of technological sophistication: database management technology, spatial analysis technology, and visualization technology. Database management technology is needed to store, retrieve, and convert large volumes of geospatial data. Spatial analysis technology is needed to make explicit a myriad of geospatial relationships among people, housing, and places. Last but not least, map visualization technology is used to view stored data. These capabilities can be made available in various ways, some of which can be combined to provide application support. HUD can assemble synergistic combinations of the GIS capabilities to increase its GIS potential by use of the following:

- **Browser** (thin client), which provides access to display only; and
- **Browser with plug-in** (thick client), which allows access to display and manipulation (Jankowski and Nyerges, 2001).

Three levels of spatial analysis each add more explicit information about geospatial relationships. First, map display enables visual analysis. Second, geometric analysis of the points, lines, and polygons can be combined with their attribute (i.e., non-spatial character) qualifiers, such as population, ethnic background, housing stock. Third, spatial-temporal analysis takes into consideration the surrounding social characteristics, such as housing stock maturity and crime levels of an area regarding their effects on public housing.

Mapping capabilities also come in a variety of levels, each level adding to the potential for visual insight into the spatial analysis performed and the database management undertaken. Map types of various forms (e.g., data magnitude by area in a choropleth[6] map or spatially dispersed data observations as in a dot map) can be used to portray complex social and economic relationships in a spatial manner. Linking maps to scatter plots or box diagrams can promote exploration of more interactive spatial relationships, and, in a sense, help to develop deeper knowledge about a housing and urban development topic.

Database technology comes in various levels of sophistication, each level adding more flexibility for retrieving and making available the stored

---

[6]Choropleth maps are divided into parts corresponding to the physical extents of the enumeration areas and these parts are shaded according to the value of a variable for that area. See < http://www.mimas.ac.uk/argus/Tutorials/CartoViz/PopViz/Choro1.html#Density> for details.

geospatial data. Three of these data management levels include the following approaches:

1. **Standalone workstation** approach to data management encourages duplication of data and duplicative efforts in processing.
2. **Local area network** approach although useful within a building to reduce duplication close by, does not foster broad-based sharing of information.
3. **Enterprise network approach** to data storage fosters data sharing, but is more expensive because it requires more coordination of data production and use.

More sophisticated GIS capabilities can be promoted for in-house HUD use and for the use of HUD's technically advanced partners. Efforts to devolve spatial analytical abilities to local users involving training and introduction of support tools are also needed. Potential in-house and advanced capabilities that HUD could develop include:

- Data aggregation for wider data distribution,
- Data aggregation functions to provide high-resolution data to users while protecting confidentiality,
- A software tool to perform data format conversion for major datasets,
- Robust GIS software platforms with multiple levels of capability to support different levels of user ability, and
- Spatial statistical modeling software.

HUD can provide web services that select, cross-reference, aggregate, document, and feed HUD-specific data into local systems. Ensuring that these web service components are interoperable, accessible, and protective of privacy are major challenges. To support web services that allow online map production, HUD would do well to track the development of the "geographic markup language"(GML)[7] and related GIS and web-service standards efforts of the Open GIS Consortium and the World Wide Web Consortium.

GIS support tools can respond to the wide range of needs and capabilities that characterize HUD's user community. Advanced tools and capabilities such as models are useful for analyzing urban development patterns and for developing housing policy.

---

[7]GML is an encoding system for geographic features that is intended to support both data storage and data transport. See the OpenGIS Consortium for details (<http://www.opengis.org/info/techno/specs/00-029/GML.html#GMLOverview>).

Simulation Models of Urban Development and Housing

Interventions in one neighborhood may affect the residents not only of that neighborhood but of other neighborhoods that are linked through the metropolitan housing and labor market. Research is aimed at developing simulation models suitable for use in analyzing these complex urban effects. The Urban Institute Model (de Leeuw and Struyk, 1975) and the NBER HUDS[8] model (Kain and Apgar, 1985) illustrate early attempts to provide simulation models for policy analysis of metropolitan housing markets. Both models attempted to simulate the effects of housing policies such as the Housing Allowance Program using a micro-simulation of the housing market. The usefulness of these models in practical policy applications was limited by the high cost of developing the models, extensive data and computational requirements, limited theoretical and statistical methods, lack of geographic detail, and limited validation of the models.

Modeling efforts have continued since the 1980s but have not been directly tied to housing policy. For example, micro-simulation models use a large sample of households to assess the effects of financial policies, often over long intervals that require "aging" persons and households in the sample over time. These models have been developed and used widely for policy analysis since Guy Orcutt developed the approach in the 1950s (Hanushek and Citro, 1991; Orcutt, 1957, 1960). Other examples of micro-simulation models include the Dynamic Simulation of Income Model (DYNASIM), used by the Urban Institute to evaluate effects of alternative social security rules on different types of families; the Cornell Simulation Model (CORSIM), and the Micro-Analysis of Transfers to Households (MATH) Model developed by Mathematica, Inc. These and other micro-simulation models for policy applications provide considerable detail about persons and households, but, as they are intended for national policy applications, they lack spatial detail that would make them useful for analysis of housing and community development policies.

Models have also been developed for metropolitan-scale land use and transportation planning, and have come into widespread use since the 1980s. The Clean Air Act Amendments of 1991, and ISTEA and TEA21[9] designate Metropolitan Planning Organizations as the principal agents in developing coordinated regional transportation plans, and require coordination of land

---

[8]National Bureau of Economic Research Harvard Urban Development Simulation (NBER HUDS).

[9]Respectively, Intermodal Surface Transportation Efficiency Act of 1991 and Transportation Equity Act for the 21st Century.

use and transportation planning processes. The DRAM/EMPAL[10] model (Putman, 1983) has been widely used for metropolitan-scale land use modeling, but lacks any representation of the housing market, and has only a modest degree of spatial detail. The MEPLAN (Echenique, 1994) and TRANUS (de la Barra, 1995) models[11] are larger-scale, metropolitan models of land use and transportation that share an approach based on spatial input-output modeling, and do include some representation of markets for real estate, though with relatively little spatial detail. The UrbanSim model has been developed recently to evaluate metropolitan transportation and growth management policies. This model uses a micro-simulation approach that makes extensive use of GIS and parcel-level data, and simulates processes of household location, business location, and real estate development and prices (Waddell, 2002).

Urban simulation models can be used to analyze the dynamics of neighborhoods and metropolitan areas and to evaluate the effects of housing and community development policies. Efforts in these areas would benefit from the lessons of early housing policy simulation efforts. Advances from ongoing development of micro-simulation and metropolitan land use and transportation models could be integrated into housing policy analysis.

Conclusion: At present, local data users including local governments and advocacy groups use GIS mostly for visualization. This allows users to view only a few variables at a time, but does not promote the application of spatial data to complex urban and community issues. While capability varies among users, the development of spatial analytical ability is important for both professional researchers and for local data users.

Recommendation: To help community groups and local governments develop spatial analysis capabilities, HUD should support the development of tools for spatial analysis. PD&R should support the development of on-line/downloadable analytical tools that incorporate multivariate techniques.

SUMMARY

HUD serves a broad spectrum of data users and stakeholders in urban and community development including some of the most disadvantaged and

---

[10]The Disaggregated Residential Allocation Model (DRAM) and the Employment Allocation Model (EMPAL).

[11]See <http://tmip.fhwa.dot.gov/clearinghouse/docs/landuse/compendium/dvrpc_toc.stm> for a review of these models.

underrepresented groups in the nation. GIS and relevant spatial data can provide HUD's users from technically sophisticated urban researchers to neighborhood advocacy groups with the ability to collect, analyze, and present data in powerful ways. HUD uses a variety of tools to disseminate and promote the analysis of these data. More progress is required in building relationships for collecting data, disseminating spatial information and know-how to HUD's clients, and devolving spatial analytical capabilities to communities.

4

Research and Policy Development

PD&R is directly involved in formulating potential urban policy solutions and monitoring and evaluating policy once it is in place. These policy issues include homelessness, fair housing, housing assistance programs, mortgages and lending, and assisting in community and economic development. PD&R is beginning to realize the potential of GIS in these areas. GIS offers multiple benefits to PD&R in terms of thinking about housing and urban issues and developing coherent public policy responses. Two of the most significant benefits are the ability to layer data from multiple sources and look at data at different scales or geographies.

HUD can use GIS to produce maps that show the distribution of public and federally assisted housing. Information about the spatial distribution of housing could show public housing authorities and other HUD field offices where their clients are and where public housing should be. Public Housing Authorities normally operate their facilities as specialized enterprises, concentrating primarily on the housing units themselves and rarely considering the surrounding framework of neighborhoods, cities, or metropolitan areas. Using GIS locally and building relationships to gather and make available data on housing and other urban conditions could inform policies that affect public housing. Data showing areas of growth in employment opportunities, public transit stops, school district data, prevalence of crime, and other themes relevant to the targeting of HUD resources could improve the agency's efficiency and effectiveness in meeting mission goals. Currently, HUD's field offices often lack both adequate data and staff who are proficient in GIS. GIS is a tool for data management and spatial analysis but the information derived from GIS is only as accurate as the data that went into the system in the first place, and as relevant as the questions posed. Understanding housing markets and the demand and supply of different types of housing is important. These gaps in data and staffing leave local HUD agencies with inadequate information for making decisions about how and where they should allocate their resources for maximum effectiveness.

Using GIS to collect, store and deliver data, and ensuring the quality of the data are important, but the application of these data to policy analysis and planning depends on the relevance of the research questions posed. In addition, the relevant data (e.g., Census of Population and Housing; American Housing Survey, Department of Labor employment data, satellite imagery, EPA air quality data, DOT traffic and accident data, airport noise exposure data) have been collected for many different applications and must be adapted if HUD's clients and partners are to use them. Data have meaning only within the context of an argument or hypothesis about how something works.

This report adopts a regional/metropolitan-level focus for addressing urban and housing issues, as described in Chapter 1. HUD can expand its research at the regional and metropolitan level to include geographic analysis of the spatial dimensions of urban poverty, the dynamics of neighborhood change, and market trends that affect the U.S. housing markets. This chapter discusses the potential of an expanded urban research agenda that is appropriate for HUD as a federal agency and identifies priorities for geographic analysis of urban and housing issues.

THE SPATIAL DIMENSIONS OF URBAN POVERTY

Understanding urban poverty requires attention to processes at the regional and metropolitan levels that result in inner-city poverty. GIS can help integrate data from multiple levels to facilitate regional analyses. The dynamics of neighborhood change and the factors that concentrate poverty in urban areas can also be analyzed using geographic data and tools. The poor are often spatially segregated from the middle class and physically removed from basic services, such as health care, childcare, and retail, and from cultural amenities, such as libraries and museums. Both the percentage of inner city neighborhoods that are poor and the percentage of poor people living in those neighborhoods have risen in recent decades (Jargowsky, 1997). Similarly, although poverty rates have declined for many groups, the income gap between the rich and the poor is widening (Lichter and Crowley, 2002). Understanding the spatial dimensions of urban poverty and neighborhood change is essential to carrying out HUD's mission.

Regional Analyses

There is growing scholarly and political recognition of the importance of regional analysis in dealing with the problems of low-income localities. For example, Orfield (1997) and Jargowsky (1997) identify processes that shape the conditions within which housing programs are situated. These processes operate at multiple geographic scales. In the past, poverty and segregation in urban housing markets were often explained in terms of locally specific analyses of individual behavior or neighborhood characteristics, without examining the processes operating at broader scales. Examples of broader or multi-scale processes that impact urban housing markets include middle-class flight to suburbs, patterns of service and high-tech industry siting, and the administrative structures of local, urban, and state government.

Regional spatial analysis provides a more comprehensive account of the problems of poor localities. Efforts toward community empowerment should address the regional processes that create the problems confronting communities and localities. Khadduri and Martin (1997) suggest that data on the positive factors affecting families should be included in analysis in addition to the negative neighborhood factors such as crime and homelessness that are often the focus of analysis. Positive factors may include accessibility, services, formal and informal support networks, and income diversity. GIS can address these multi-scalar questions relevant to urban poverty in terms of these broader forces and processes that shape neighborhoods. Regional spatial analyses of this kind are not simple or user-friendly (Luc Anselin, University of Illinois, Urbana-Champaign, personal communication, 2002); rather they require a trained workforce. Box 4.1 presents an example of sophisticated statistical analysis that supports anti-discrimination efforts.

The committee suggests the following research questions that lend themselves to spatial analysis and may illuminate the relationships between urban and suburban processes, and housing market conditions and trends within low-income communities. Addressing these issues using spatial data requires experience and expertise in geographic analysis and research.

- To what extent do policies that promote or allow rapid urban decentralization contribute to decline in inner-city neighborhoods?
- How effective are place-based investments in inner-city neighborhoods if rapid decentralization undermines inner-city housing markets? What are the alternative policies?
- How effective are growth management policies such as urban growth boundaries and smart growth initiatives in curbing the rate of urban expansion?

- Do growth management policies contribute significantly to housing affordability problems?
- Is there a threshold on the size of the region for metropolitan governance, beyond which regionalism may not work? How can we integrate "things-regionalism" with "people-regionalism"[1] in urban and housing development?
- What is known about urban processes such as sprawl, economic polarization, and population growth; and the interactions among these and other processes at global, national, state, regional, metropolitan, and local scales?

GIS and Section 8 Housing Policy Analysis

This section describes the uses of GIS in analyzing Section 8 housing policy (Box 4.2). The examination of the application of GIS to Section 8 housing issues illustrates the importance of research on the spatial dimensions of urban poverty, the role of data intermediaries, and data issues including privacy concerns and the determination of causality. Using GIS for Section 8 Housing Policy Analysis encompasses understanding the needs and concerns of the residents that can be addressed using GIS, responding to those needs, keeping data on what worked and what did not, and exercising judgment about what should be done next.

Policy research that can be addressed using GIS include questions about the concentration of people and assisted housing; and the concentration of HUD Section 8 tenant-based assistance program households and employment opportunities (Thompson and Sherwood, 1999). Many internal HUD policy analysts, as well as external HUD research partners, are interested in population and income distribution and in the need for housing and services at neighborhood to regional scales. Thompson and Sherwood (1999) developed a guidebook for GIS use in which many of the examples focus on data for Section 8 housing. The guidebook presents a method for using micro-data (individual households) and a corresponding method for addressing confidentiality concerns.

Confidentiality may be a concern when collecting and disseminating data on the spatial distribution of low-income housing. Methods should be devised

---

[1]"Things-regionalism" refers to the most common local government attempts at regional initiatives that focus on a single function such as transportation, watersheds, sewers, and emergency management. The most extreme poverty in America is typically geographically concentrated, suggesting a need for "people-regionalism" to promote diversity, balance, and stability in every area of a region (Cisneros, 1996).

to avoid identification of individual households but still provide high-resolution data that addresses policy concerns. Methods involve deletions and geocoding procedures; restricted access to data; rounding numbers into larger groups, aggregation units, and scale; and random displacement of certain point data. Determining causality is another difficulty. GIS data analysis, like other data analyses, cannot compute causality. It may take many maps to explore a topic before an analyst may gain insight into relationships among housing variables.

A robust data management approach is fundamental to having access to data for mapping purposes because multiple maps are often necessary to develop an "analysis scenario." Figure 4.3 depicts HUD-assisted housing relative to employment concentrations mapped by traffic analysis zones (TAZs)—the units most often used to compile employment data for transportation purposes. Figure 4.4 shows the TAZ employment data aggregated to concentrations.

Although maps can be used to answer many questions, they can also prompt questions such as: Do people in Section 8 housing get higher paying jobs first, then move? Or, do people move and then get higher paying jobs? Is this pattern a result of household mobility? GIS can inform questions and guide next steps for research on the distribution of Section 8 housing allocation in relation to employment.

Dynamics of Neighborhood Change

The process of neighborhood change has been a subject of academic research in several disciplines for many decades, and yet significant gaps persist in our understanding of the dynamic processes that produce decline, revitalization, gentrification, and other urban processes. Since the early twentieth century, researchers in various disciplines have studied neighborhoods. Early examples include application of the ecological lens of invasion and succession to neighborhood studies (Park, 1926), analysis of economic and social factors that contribute to neighborhood decline and revitalization (Downs, 1981), and theories of tipping behavior [2] (Schelling, 1978). Sociologists have explored neighborhood patterns of racial segregation (Farley and Frey, 1994; Massey, 1990) and the emergence of concentrated poverty

---

[2] A model explaining that the collective action of individuals may produce segregation even when the individuals prefer integration. Tipping behavior is the racial make-up of a neighborhood that prompts flight from the neighborhood (Schelling, 1971, 1972, 1978).

(Jargowsky, 1997; Wilson, 1987); geographers have studied spatial patterns of gentrification (Smith, 1996).

In spite of this sustained and broad spectrum of social science research on neighborhoods, a coherent synthesis and policy responsive to the dynamics of neighborhoods—especially poor neighborhoods—remains elusive. Among the pressing questions that remain are:

- What internal and external factors, in varying combinations, influence the decline and revitalization of neighborhoods?
- How do the mechanisms of decline and revitalization work, and how susceptible are they to intervention?
- How are expectations of neighborhood residents and outsiders about the future of a neighborhood formed and changed?
- How effective are different forms of intervention, under varying market conditions, at stemming or reversing neighborhood decline?
- Is gentrification becoming more widespread; if so, under what conditions, and how can displacement of low-income residents be mitigated in gentrifying neighborhoods?

BOX 4.1
Race and Mortgage Lending

Race reporting is required from mortgage lenders as a result of the Home Mortgage Disclosure Act,[1] which monitors lending practices in minority communities, however, racial disclosure to lenders when borrowing by phone or Internet is not required. Subsequently, the second largest racial/ethnic group of those seeking mortgage credit in the United States is listed as "Not Reported." A recent study set out to analyze the geographic expression and causes of the "Not Reported" racial/ethnic designation. To this end, a GIS-enabled spatial analysis of nondisclosure reporting in Atlanta, Georgia, was conducted using the following three econometric models demonstrating:

- The differences in individual characteristics between disclosing and non-disclosing loan applicants;
- The degree to which non-reporting of racial/ethnic identity results from institutional factors rather than deficiencies of loan applications; and

---

[1]The Home Mortgage Disclosure Act (HMDA), enacted by Congress in 1975 and implemented by the Federal Reserve Board's Regulation C, requires lending institutions to report public loan data.

- The influence of neighborhood-level characteristics on nondisclosure rates after controlling for both individual and institutional measures.

The study found that the geography of nondisclosure coincided with the areas that lacked accurate data for lending research and had a high proportion of African-American neighborhoods. Based on these findings, the study concludes that there is a need for coordinating outreach efforts to publicize the importance of reporting race/ethnicity data to enforce civil rights.

SOURCE: Wyly and Holloway, 2002.

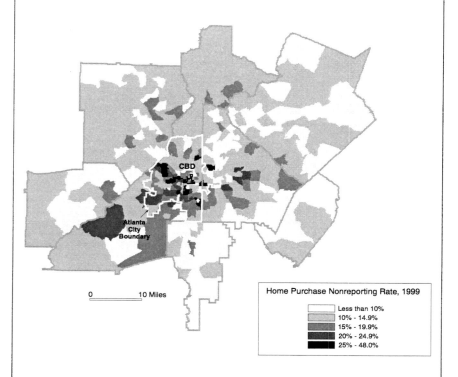

Home Purchase Nonreporting Rate, 1999

Less than 10%
10% - 14.9%
15% - 19.9%
20% - 24.9%
25% - 48.0%

0   10 Miles

FIGURE 4.1   Share of home mortgage applications in Atlanta, Georgia without race-ethnicity information, 1999. Pattern confirms that nondisclosure rates are highest in predominantly African-American neighborhoods. SOURCE: Wyly and Holloway, 2002.

BOX 4.2
GIS and Section 8 Housing Choice

The Section 8 tenant-based housing assistance program provides subsidies that allow low-income families to live in higher-quality private-market rental housing. This approach aims to more closely match housing preference to provision, and to increase opportunity beyond what is typically available near public housing, which is frequently located in high-poverty neighborhoods. Current research shows that some Section 8 families live in poorer and racially segregated neighborhoods (Turner et al., 1999). One possible cause of this concentration is the lack of sufficient counseling about rental housing. A recent study presented a prototype application, the Housing Relocation Assistant (HRA), which uses GIS to display neighborhood characteristics for the selection of Section 8 rentals based on user preferences. The prototype uses seven categories of objective indicators across the metropolitan area of Pittsburgh, Pennsylvania, for multi-criteria analysis. These include:

- Availability of high-level entry-level employment
- Availability of affordable housing
- Public transit accessibility to jobs
- Social services support
- Quality of education
- Public safety
- Local amenities and demographic characteristics
- Quality of life

The HRA prototype contains more specific indicators of these seven broad categories. Most of these data were already widely available from local, state, and federal sources. The role of the housing counselor would be key to bringing such a prototype into practice and to assisting the Section 8 family to determine the best indicators for their decision. When the criteria are entered into this system, the local areas are correlated to these criteria based on user preference, and alternative destinations can be ranked by these preferences. GIS allows the information to be displayed visually and functions as a decision-support tool. (See Figure 4.2.)

SOURCE: Johnson, 2002.

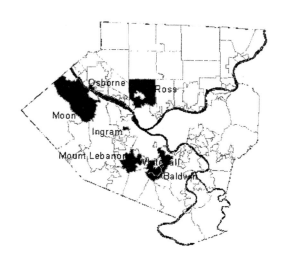

FIGURE 4.2 Map of Allegheny County, Pennsylvania municipalities. Relocation areas that meet the hypothetical user criteria are marked in gray. SOURCE: Johnson, 2002.

Poverty Concentration and Racial Segregation

It is said that rising tides lift all boats, but there are clear winners and losers in the economic boom of the late 1990s. Research is needed to address the underlying causal mechanisms and consequences of the economic and social factors that result in poverty, segregation, homelessness and other urban ills. Box 4.2 provides an example of how GIS can be used to analyze poverty concentration. Data from the 2000 Census may stimulate new research to document trends in poverty concentration and racial segregation.

Much research has been done on patterns of poverty concentration and racial segregation, but many questions remain unanswered or inconclusive, for example, the role of socio-economic class versus race in determining segregation patterns. New multiple-scale segregation measures using GIS (Wu and Sui, 2001), as well as theoretical perspectives such as Sen's entitlement theory (Sen, 1976), can be applied to these efforts to explain the housing situation of the urban poor. In addition, the growing urban digital divide—the existence of the information-rich and the information-poor—should be taken into account.

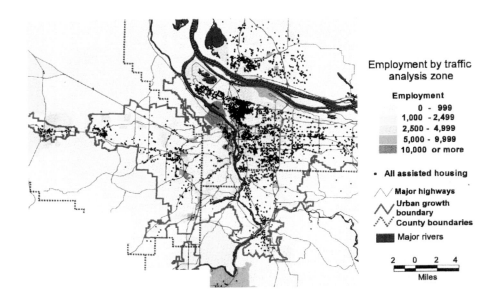

FIGURE 4.3   HUD-assisted housing relative to the location of employment, 1996, Portland (OR-WA) MSA. Assisted housing includes all types of HUD-assisted housing. SOURCE: Thompson and Sherwood, 1999.

Neighborhood Effects

Literature is extensive on the effects of neighborhood conditions in poverty-ridden, segregated communities on individual social outcomes. Research sheds light on the effects of neighborhood concentration of poverty on teenage pregnancy and dropping out of school (Crane, 1991), and on a range of other outcomes, such as crime, cognitive skills, and labor market success (see Jencks and Mayer, 1990 for a review). In large part, this research has helped to motivate and shape HUD's Moving to Opportunity Program,[3] a 10-year research demonstration project providing tenant-based rental assistance and housing counseling to assist very low-income families to relocate from poverty-stricken urban areas to less poor neighborhoods. It has also aided the

---

[3]The final report of this study, *Families in Transition: A Qualitative Analysis of the MTO Experience*, is available on-line at <http://www.huduser.org/publications/pdf/mtoqualf.pdf>.

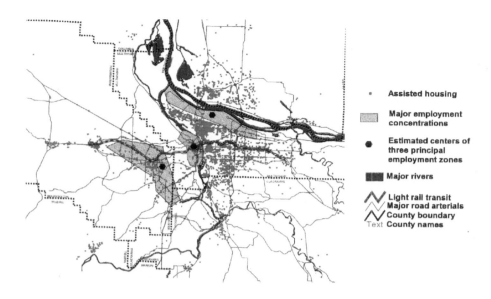

Legend:
- Assisted housing
- Major employment concentrations
- Estimated centers of three principal employment zones
- Major rivers
- Light rail transit
- Major road arterials
- County boundary
- Text County names

FIGURE 4.4 HUD-assisted housing relative to major employment concentrations, Portland (OR-WA) MSA Assisted housing includes all types of HUD assisted housing; Employment concentrations were demarcated from maps of employment distribution by traffic analysis zones. SOURCE: Thompson and Sherwood, 1999.

Gatreaux project, which provided portable vouchers, mobility counseling, and housing location assistance to 7,100 families residing in Chicago's public housing to assist their moving to private housing, mostly in Chicago's suburbs. Moving to Opportunity and the Gatreaux project have been closely studied in an attempt to document the effects of moving public housing residents into suburban environments (Rubinowitz and Rosenbaum, 2000).

Isolating neighborhood effects statistically from individual and broader economic and social context effects has proven difficult. Although the Panel Survey of Income Dynamics[4] provides a valuable national source of information for longitudinal analysis of the socioeconomic conditions of persons and households, the sampling frame for the survey was not designed for the analysis of neighborhood effects, and geographic clustering within the survey is insufficient to identify neighborhood effects. A major new data initiative

[4]<http://www.isr.umich.edu/src/psid/>.

has been proposed to support research to disentangle neighborhood from individual effects. It would provide a sampling of young adults within neighborhoods stratified by neighborhood resources, school district resources, and political jurisdiction resources, and it would oversample poorer neighborhoods (Paul Jargowsky, University of Texas at Dallas, personal communication, 2002). A retrospective history of addresses would be connected to neighborhood characteristics, and would follow its subjects longitudinally, including children of respondents. Not only could neighborhood effects be separated from individual effects, but also the effects of higher levels of social and economic context, such as school and city setting, could be identified.

The policy implications of unraveling the effects of neighborhood conditions are considerable. Data on the effects of neighborhood transitions can inform the development of policies such as Moving to Opportunity, housing voucher programs, and also policies aimed at improving neighborhood conditions and assisting community planning which are among HUD's goals. Recent research on neighborhood effects includes Ellen (1999), Ellen et al., (2001), and Schill and Daniels (2002).

Statistical analysis tools to address the problems presented by spatial data at the neighborhood and other levels have found only limited application within the realm of urban development and housing policy. Examples of innovations in statistical spatial analysis[5] that could augment and support a broad research agenda on housing and urban issues and promote HUD's mission and goals include:

- Spatial pattern analysis and interpolation using statistics,
- Spatial econometric modeling,
- Multi-level, or hierarchical linear modeling, across-scale analysis,
- Bayesian statistical models of spatial and temporal patterns and processes,
- Spatial metrics,
- Spatially weighted regression,
- Data mining and knowledge discovery techniques, and
- Analysis of housing and labor market interactions and spatial forecasting.

Conclusion: As a federal agency with a strategic goal to provide a decent, safe, and sanitary home and equal opportunity for every American, creating a vision for the future of urban America is an appropriate endeavor for HUD. This implies a broad urban research agenda.

---

[5]For descriptions of spatial statistical tools and methods see <http://www.statistical.org/>.

Recommendation: HUD should expand its research portfolio to emphasize the following urban issues:

- The spatial dimensions of urban poverty in the United States;
- The dynamics of neighborhood change; and
- Market trends that affect the U.S. housing market.

MONITORING HOUSING MARKET CONDITIONS AND TRENDS

Among the goals of PD&R is "ensuring the availability and accuracy of essential data on housing market conditions and trends." Understanding these conditions and trends is crucial to policy development. A low-income housing subsidy in a neighborhood that is experiencing steady attrition of middle-income residents to other neighborhoods provides an example. Such a policy would add to the low-income housing building up in the neighborhood and would do nothing to stem the exodus of middle-income households and the resulting concentration of poverty in the metropolitan area.

Housing conditions and trends should be analyzed over several scales, because conditions at one scale may significantly alter the effects of an intervention at another scale. Neighborhood conditions are influenced heavily by the metropolitan context in which they exist; for example, housing market conditions in newer suburban communities interact with conditions in older inner-city and inner-ring suburbs. Without accounting adequately for these types of interactions, investments and policies may be ineffective or counterproductive. GIS can be a useful tool to analyze multi-scalar processes for policy development and investment choices (Box 4.3).

Metropolitan Markets and Sub-markets

Housing markets are often defined and monitored at metropolitan and sub-metropolitan scales. Metropolitan areas form natural market areas for housing and labor. A system for monitoring economic and housing conditions should include a metropolitan scale component. National conditions that affect metropolitan areas are tracked by the American Housing Survey and by national compilations of construction starts, sales prices, and rents. Local conditions within specific metropolitan markets are most salient to HUD data users but local area conditions can deviate dramatically from national trends due to:

- Regional variation in the timing of business cycles;
- Local economic structures;

- Variations in inter-regional migration streams that influence demand for housing; and
- Local policies, for example, zoning and land use regulations that affect the responsiveness of local housing supply to changes in market.

BOX 4.3
Application of GIS to the
Assessment of Housing Conditions in Dallas, Texas

In 1993, the city of Dallas Department of Housing and Neighborhood Services commissioned a study of housing conditions within the city. The objectives were to assess the physical condition of the city's housing stock; to estimate the costs required to bring all substandard housing to standard condition; to describe the demographic characteristics of the population; and to provide reliable, empirical data upon which housing programs and strategies could be designed. The city wanted to be able to identify the geographic areas in which substandard housing was concentrated and in which substandard housing was likely to increase.

The data collection effort included surveys (a reconnaissance survey; a household survey; and mail, telephone, and in-person surveys) and physical inspections by trained building inspectors of a subsample of housing units. The purpose of the physical inspection was to establish statistical models to predict renovation costs for the full sample and its expansion to the city's housing stock.

Costs to bring substandard housing up to standard condition were estimated via regression analysis. The final set of independent variables used in the development of the cost estimates included: deterioration ratings, a neighborhood deterioration score, a dummy variable for frame construction, age of building, and number of violent crimes. The adjusted R^2 was 0.59, and the coefficients were statistically significant at the 5 percent confidence level.

Using the regression model, and applying the model to the full sample and then to the full housing stock of the city, at a parcel level, cost estimates for renovation were generated. The predicted cost for bringing the entire housing stock of the city of Dallas up to the HUD standard was estimated as just over $900 million, representing slightly less than 4 percent of the total residential property value of the city in 1993. The costs were disproportionately represented by single-family structures for which renovation costs were estimated as $783 million, or 4.4 percent of single-family appraised value for the city as a whole. Multi-family housing renovation would cost $124 million, or 2.4 percent of the appraised value of multi-family structures in Dallas.

By integrating spatial analysis and visualization using GIS techniques, this analysis of housing conditions was carried out at a parcel level of detail. Figures 4.5 and 4.6 depict the predicted conditions of the single family and multi-family housing stocks, at a parcel level, and reveal spatial clustering that warrants closer attention by the city. By using the predictive models estimated as part of the study, administrative records can be systematically monitored to provide leading indicators of emerging potential problems with the housing stock, in a timely and cost-effective way that facilitates timely intervention.

SOURCE: Waddell, 1994.

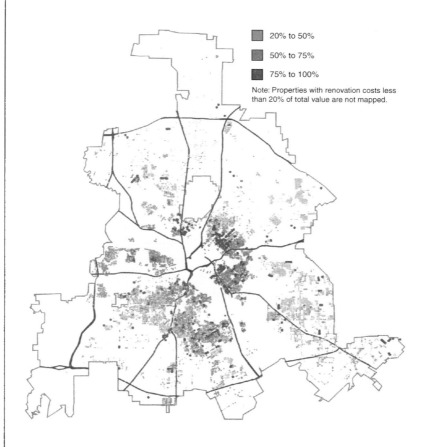

20% to 50%

50% to 75%

75% to 100%

Note: Properties with renovation costs less than 20% of total value are not mapped.

FIGURE 4.5   Estimated renovation cost as a percent of property value for single-family housing, City of Dallas, 1993. SOURCE: Waddell, 1994.

BOX 4.3 Continued

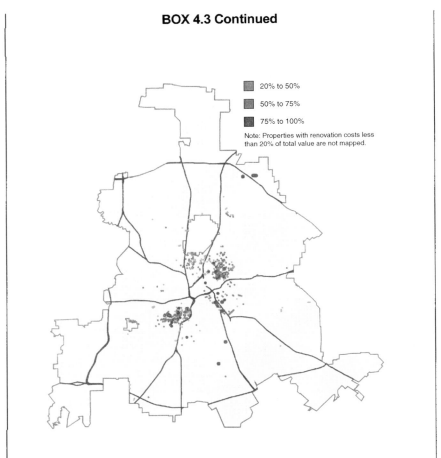

20% to 50%

50% to 75%

75% to 100%

Note: Properties with renovation costs less than 20% of total value are not mapped.

FIGURE 4.6   Estimated renovation cost as a percentage of property value for multi-family housing, City of Dallas, 1993. SOURCE: Waddell, 1994.

A systematic effort on the part of PD&R to monitor metropolitan housing market conditions and trends would, in part, duplicate monitoring activities by private real estate market research firms, at least in many metropolitan areas. Real estate tracking of market conditions for the metropolitan areas and for sub-markets within the area can be defined by housing type and geographic area. These firms keep up-to-date detailed information on vacancy rates, rents, sales prices, and new construction; and issue regular reports on these conditions through local media and/or private reports available to subscribers. HUD could develop cost-effective and reasonably accurate monitoring systems by partnering with market research firms, or in some cases non-profit agencies or university research centers that deal with real estate and community planning. HUD could use GIS to integrate local, metropolitan/ regional-level, and national-level data to analyze the effect of ordinances such as minimum lot size, bans on multi-family housing, and other zoning ordinances on the cost and supply of housing in suburban areas.

Key Indicators of Local Housing Market Conditions

A monitoring system relies on key indicators that efficiently describe the most salient characteristics of the local housing markets. At a minimum, these metropolitan housing market indicators would include:

- Vacancy rates by type of housing,
- Sales prices and rents by type of housing,
- New construction, conversion, and demolition of units by type,
- Drivers of housing market conditions and trends including:

    o Employment change by sector,
    o Changes in wages and their distribution by sector and occupation,
    o Population change by type,
    o Interest rates, costs of construction, tax policies.

- Cost burdens for renters and owners at various income levels.

Sub-metropolitan spatial detail (resolution) is vital to understanding the dynamics of metropolitan housing markets. Some form of sub-market definition that differentiates by type of housing and by geographic market area will significantly enhance the usefulness of a monitoring system for housing conditions and trends.

Hedonic Analysis of Quality-Controlled Prices

In monitoring sales prices or rents, the typical approach is to report median prices and rents from recent transactions, and compare them with the preceding month or with the same month in the previous year. This approach provides useful information, but does not adequately control for the changing composition of the housing stock reflected in the transactions, which might account in large part for observed changes in prices. In other words, median prices may rise because larger houses (or better quality houses, or those in more desirable locations) are being built or are sold as market conditions change, rendering the median transaction price a potentially misleading indicator of housing prices.

A more effective indicator of prices or rents is the use of a hedonic, or quality-controlled, housing price index (Waddell et al., 1993). Hedonic analysis examines prices and demand for individual sources of pleasure so that it can used to understand the relationship between quality and price. Such analysis can be used to analyze housing markets and to understand the relationships among rental prices and the characteristics of structures, such as size, number of bedrooms, lot size, and quality of construction, in addition to the locational characteristics that so influence housing values. Using hedonic analysis, trends in price over time can be estimated in a way that holds constant the other factors in the analysis.

Many of the factors that affect housing markets are spatial. It is said of real estate that what matters most is location. A significant advantage of hedonic analysis is that it can readily incorporate substantial spatial detail, allowing analysis of how various locational factors influence the housing market. Spatial factors include location of public housing, concentration of poverty, crime patterns, access to transportation, and amenities (e.g., parks, supermarkets, libraries) and disamenities (e.g., waste sites, heavy industry, abandoned parcels) that vary spatially. GIS can be used to carry out analysis of this nature.

Analyzing Housing Market Conditions and Trends

Although it is clear that monitoring housing market conditions and trends is important to improving the ability to make informed and effective policy choices, there remains a significant gap between useful market information and inferences about the influence of the market on specific policy choices. Valid analytical methods are needed to make such inferences and carry out such analyses.

HUD can use GIS-supported analysis agency-wide for program monitoring and oversight including self-assessment and continuous improvement, internal quality assurance procedures, customer satisfaction, community and resident involvement, and the cost-effectiveness and affordability of programs as recommended in the NAPA report on HUD-assisted public housing programs (McDowell, 2001).

Specifically, research is needed to develop analytical methods to allow HUD and its constituents to:

- Use results from monitoring of housing conditions and trends within particular metropolitan areas to infer potential effects on low income communities and residents;
- Assess the effects of these conditions and trends on the viability of specific HUD or local community investments and policies;
- Assess, in turn, the effects of HUD or community investments and policies on these conditions and trends; and
- Evaluate the effects of HUD or community investments on specific outcome measures of interest (e.g., segregation, crime, or community involvement such as voluntary associates per capita or per capita hours of participation in community-based activities [NRC, 2002c]), holding constant the effects of background market conditions and trends.

Conclusion: HUD needs internal spatial analysis capabilities and a systematic approach to monitoring metropolitan housing market conditions and trends in order to help local governments and nongovernmental groups develop their policy analysis capabilities.

Recommendation: To monitor and analyze metropolitan housing market conditions and trends, HUD should:

- **Identify and adopt means and formats for routine collection of housing-related data relevant to user needs and agency mission goals at regular intervals, along with development and adoption of a standardized method for data analysis; and**
- **Perform research towards the development of spatial analytic tools to address quality-controlled price indices and variations in local context, and for time-series and comparative analyses between and among places.**

PRIORITIES FOR GEOGRAPHIC ANALYSIS OF URBAN AND HOUSING ISSUES

Spatial analysis of urban poverty, neighborhood dynamics, and market trends affecting the housing market will provide data and information that HUD and its clients need to address housing and urban issues. In this section, the committee outlines research priorities for HUD in the agency's efforts to use geographic information to address urban and housing issues. The committee also offers ideas for research in support of national security needs and trends in communications and other technology development that are not among the research priorities but are issues that may affect urban development. To provide strategic guidance for the agency, the committee underscores the importance of an agency-wide GIS and identifies research areas that will contribute to the HUD's internal management and program assessment, and promote the agency's broader mission goals.

Conclusion: An agency-wide GIS can be used to examine urban issues and housing trends across multiple geographic scales from neighborhood to region, and at different levels of spatial resolution in a metropolitan, regional, or international context. HUD can work with internal datasets and with those produced by partners; investigate the spatial structures and social processes at work in a metropolitan or regional context that underpin many community concerns with housing and investment; and engender participation among partners with interests in policy analysis, research, and community building.

Recommendation: HUD should incorporate into their research agenda and prioritize spatial analysis of the following urban issues at the regional and metropolitan-level:

- **Housing market conditions and trends,**
- **Effects of these conditions on HUD program design and implementation,**
- **HUD program effectiveness and effects on communities,**
- **Interactions among communities in metropolitan areas,**
- **Dynamics of neighborhood change including poverty concentration, racial segregation, and neighborhood effects, and**
- **Housing and labor market interactions including regional and cross-border analyses.**

Urban data are useful for multiple applications. Table 4.1 presents examples of current work and opportunities for the application of geographic information using GIS at HUD. The entries in the table are potential GIS applications for HUD that require GIS tools for support.

HUD and Homeland Security

Since HUD has custody of much valuable detailed information related to communities across the country, the committee believes that the agency can play an important role in various new federal initiatives dealing with homeland security. Issues in which HUD can play an important role using GIS include efforts to:

- Understand the particular vulnerability of highly populated urban areas with important government, commercial, military, or historic buildings; develop better methods of identifying the most vulnerable communities; and prepare plans to respond to the potential attacks;
- Understand the role of GIS in the context of homeland security issues, for example, research on regional processes including border issues and trans-border process such as those at work in the colonias along the U.S.–Mexico border; and
- Ensure that local-level data are available for integration in homeland security's information base.

Effect of Technological Innovations on Housing and Urban Development

Technological innovations, especially in telecommunication, play crucial roles in determining the trajectories of metropolitan development. Despite the growing interest among scholars in this topic, there have been no systematic federal initiatives focusing on the issue since the publication of *The Technological Reshaping of Metropolitan America* by the now defunct Congressional Office of Technological Assessment in 1995. Yet, new technological innovations in telecommunication (mobile phone, wireless communication) and computer technology during the past 5 years were unprecedented and their effects on society in general, and housing and urban development in particular are still unknown. The committee offers the following suggestions for research questions:

- How will mobile phone and wireless communication shape metropolitan development in the United States?
- Although there is a huge literature on telecommuting, we know little of the consequences of telecommuting. What are the effects of telecommuting on urban development?
- To what extent is technological innovation responsible for the growing gap between information-rich and the information-poor? What role should HUD play in narrowing the growing digital divide?
- Are technological innovations moving us closer to the goals of urban sustainable development or further away from them? How can we make future urban and housing development more energy efficient and less material intensive?

TABLE 4.1 Cross-Reference HUD Responsibilities and Potential Application Categories

Potential GIS Application Categories	Conduct research for policy development	Use research to support program offices	Collect and Distribute Data to HUD and Partners
Home ownership for low-income and minorities	Study impact of housing policies; home mortgage discrimination; mortgage distribution; home ownership and demographic information for determination of fair-market rent and purchase price.	Support research needs for: Implementing the Blueprint for American Dream; Distributing FHA and other HUD-insured lending programs; and Identifying potential Home Ownership Zones.	Provide data on neighborhood characteristics including schools, transportation, access to health care to allow more informed choice for disadvantaged groups.
Colonias on Southwest U.S. border	Determine the extent of Colonias' development and associated lack of infrastructure	Inform decisions addressing: Allocation of CDBGs; Affordable housing needs in the Southwest region; and Needs of Colonias field offices.	Link state and local data to info from aerial photographs. The program needs to include data from other agencies such as DOT, HHS, EPA and Census.
Growth management	Promote regional and metropolitan land supply GIS database development	Assist communities with fiscal policy based on multi-scale analysis; Guide CDBGs to counter sprawl; and Support state-wide programs with geographic information to evaluate opportunities for affordable housing developers.	Promote inter-organizational and public participation in land use decision making, collect and maintain new building permit data.

Data integration	Create comparable data for comparative analysis	Bring practitioners, policy makers, and citizens to same page through maps when planning Urban Empowerment Zones.	Provide Web-based GIS as data integration engine
Analysis of Neighborhood Change	Identify spatial statistical relationship between and among variables	Support the research needs of fair housing assistance programs; Determine spatial concentration of Section 8 voucher recipients; and Evaluate Hope VI effectiveness.	Integrate parcel-level data from multiple metropolitan areas for comparative analysis
Public Housing (Low income and minority)	Perform multi-scalar analysis to understand neighborhood change	Inform decisions about siting subsidized housing; Identify Section 8 housing options within vicinity; and Enhance PHA performance assessments.	Make data available at finest resolution
Web site as Data Warehouse	Provide information to broaden the range of choice in residential location for low-income groups	Provide common data formats; and Accommodate vertical and horizontal data sharing.	Make it easier to maintain and provide data as needed
Partnership Building	Broaden the agencies research agenda through collaboration with other agencies	Identify spatial barriers to employment for residents of subsidized housing.	Share data among federal, state, local groups, promote data interoperability

Cell entries represent potential GIS application topics.

SUMMARY

To better understand regional- and metropolitan-level urban processes and the urban housing market, HUD should spatially enable their research portfolio by incorporating GIS into HUD research across the agency. Support is essential for local governments and other users at the local level to develop capability in spatial analysis. Programs and tools such as an online clearinghouse for spatial data research and urban simulation models using GIS and spatial analysis will promote analysis of complex urban issues that span geographic scales of neighborhood, community, region, state and the nation. Addressing these recommendations will necessitate resources including expertise in GIS, spatial analysis, geographic research, algorithm development, and spatial data manipulation.

5

The Role of Partnerships

The complexity of housing and urban issues necessitates wide-ranging interagency partnerships if HUD is to share relevant data, to identify emerging issues, and to develop policies for response and intervention. Partnerships can improve data quality, facilitate information gathering and data sharing across scales, and make data resources available for researchers to develop more sophisticated models of housing and urban issues. HUD's Office of Policy, Development, and Research (PD&R) has a goal of working through interagency groups to address housing and urban issues, and PD&R is in the forefront of HUD's efforts to use GIS to achieve this goal.

HUD produces and disseminates data on housing and housing conditions, but most are point data that tell us little about the neighborhoods and neighborhood characteristics, cities, metropolitan areas, and regions in which these data points are situated. In addition, there is uncertainty about the accuracy of these data sets, as a result of factors including difficulties with geo-coding, failure to maintain data standards, and poor program reporting.

To broaden the dialog and achieve consensus on housing and urban issues, HUD needs strong internal and external relationships, horizontally with other federal agencies and vertically with partners from the local to the national level. Table 2.1 shows the diverse responsibilities of the many partners in the Federal Geographic Data Committee (FGDC) for providing data to the nation. HUD's vertical relationships include citizens and community groups, local organizations such as city governments, agency employees, and the urban planning and research community. This chapter discusses the array of partnerships HUD has with other agencies at various levels, and makes recommendations for the maintenance and expansion of these relationships for the mutual benefit of HUD and its partners.

BUILDING INTERAGENCY PARTNERSHIPS TO SHARE DATA

Within HUD, GIS provides a framework for organization across the agency's many programs and for communication among the geographically decentralized field offices. The staffs of more than 3,400 public housing authorities, which are primarily responsible to the cities in which they are situated, also work to carry out HUD's goals. Centrally, HUD maintains a variety of databases, primarily for program management (Chapter 2). The data could be centralized and made more accessible to these groups to improve HUD's program performance through assessment and to better address housing and urban issues.

HUD can use GIS to facilitate the agency's efforts to interact with organizations beyond its institutional boundaries to build vertical and horizontal networks to share data, discuss housing and urban issues, and ultimately create public policy to respond to these issues. In this way, HUD can contribute to national data initiatives and carry out its mission to improve the accessibility and affordability of housing and contribute to urban development.

GIS can help HUD engage both communities and a variety of other actors in discussions about local and national urban public policy. Maps are very effective tools for communicating information and fostering debate on critical issues. Maps have the power to inform local planning, engage and empower community residents and organizations, promote data sharing and interagency coordination, and support public policy development and implementation.

It is significant that HUD is one of the few federal agencies working directly with cities, communities, and neighborhoods. Because of these relationships, the agency can promote public participation in decision making, narrow the divide that prevents disadvantaged communities from participating in urban and housing policy setting, and bring the capabilities of GIS to bear on issues of local and national relevance.

The following section discusses the range of HUD's relationships with other organizations at local, regional, and national levels and how the use of GIS could influence those relationships.

HUD'S PARTNERS AND RELATIONSHIPS

HUD's structure and mission necessitate a wide range of relationships among the many partners that support community development. The agency interacts directly with public and private entities from the local to the national scale. In some instances, these relationships are dictated by statutory requirements; others are voluntary relationships. Even within these general categories, relationships may be long-term or ad-hoc. HUD's relationships with communities and local agencies are very important. For example, lessons learned from a fair housing demonstration program describe patterns of residential mobility and outcomes of families participating in a program aimed at upward mobility of low-income families. The findings of that demonstration program underscore the importance of cooperation and commitment of local agencies including counseling agencies, and provision of related information and referral services in the success of such programs (Goering et al., 1999).

Relationships Within HUD

HUD is an organization of 9,000 employees with a complex structure involving a central office in Washington, D.C., and more than 80 field offices throughout the United States. It is the field offices that deal most directly with HUD's potential and actual grant recipients. Because of the spatial quality of HUD's investments, the field offices can use GIS in their work of building relationships and empowering communities.

GIS can help HUD develop and disseminate new approaches including policies to dealing with the housing conditions of the very poor. Communities are employing place-based strategies for managing growth and GIS can be instrumental in the development, dissemination, and replication of these innovative local, community-based solutions across the nation (NRC, 2002c). Many local governments at the city and county level have developed sophisticated GIS-based data systems for planning, community development, economic development, and housing. HUD can function as a clearinghouse of information including lessons learned and best practices in the use of GIS. Within HUD, GIS could also be used to test the effectiveness of the voucher program and to define the appropriate geographical areas for various government programs (see Box 5.1).

BOX 5.1
Concentration of Poverty and GIS

HOPE VI and Section 8 vouchers are two HUD programs aimed specifically at dispersing poverty. HOPE VI grants pay for the demolition of public housing units, or rehabilitation of these units, the building of new units, relocation of units, and service provision for the community. The program favors public-private partnerships in these ventures. Critics point out that under HOPE VI more than 115,000 public housing units have been demolished but only 66,000 have been or will be replaced, and many of new units are not available to the previous low-income residents. The program's intent is to serve displaced residents with the Section 8 voucher program through which HUD supplements the rent of low-income residents to provide access to better housing.

An Urban Institute study documents that HOPE VI residents relocating with Section 8 vouchers are indeed moving to less distressed neighborhoods and that the de-concentration of poverty achieved by this program since its inception in 1993 is substantial (Kingsley et al., 2001). But the report goes on to ask whether the distribution of relocated residents exacerbates or counters the clustering problems that result from the Section 8 program and suggests that important knowledge gaps must be overcome before the HOPE VI program can be fully assessed.

The replacement of Chicago's Cabrini-Green public housing complexes is a famous example of the HOPE VI program in practice. HUD committed to $50 million in HOPE VI Urban Revitalization Demonstration funds and $19 million in Public Housing Development funds to build 493 replacement public housing units. Slated for construction are 2,000 new mixed-income housing units (row housing, duplexes, and mid-rise buildings), but only 30 percent of the new housing is reserved for previous Cabrini-Green residents. Although the new mixed housing will offer a commercial district, new schools, and more integrated demographics for 30 percent of current residents, 70 percent of Cabrini-Green residents will be displaced in other parts of the city.[1] Assessing the regional scale effects of this displacement of thousands throughout Chicago will be facilitated by spatial analysis, the application of GIS, and maintenance of geo-coded address records.

1See <http://pubweb.northwestern.edu/~smc365/final/ final.html> for details.

Statutory Relationships with Local Agencies

HUD exercises statutory authority over agencies that receive HUD funding (grantees) across the nation. Grantees should provide follow-up information and progress reports to maintain HUD monetary awards. HUD can influence these relationships and gain the cooperation of grantee organizations. Grantees are largely local-level organizations. Consequently, HUD has a singular influence at the national level to mandate that collection of local data be consistent with FGDC standards for inclusion in vital national data structures, such as the National Spatial Data Infrastructure (NSDI) and Homeland Security databases.

Local and Metropolitan Partners

Local actors are significant sources of spatial data. Datasets that communities collect and maintain locally include building permits, demolitions, property assessment, code violations, and vacant and abandoned properties. Local knowledge is often superior to knowledge derived from national datasets (Barndt, 2002). People in their neighborhoods know about the context and networks that lie behind the points on a map. They know what the vacant land is used for, whether it is a playground, a community garden, or not really a vacant lot at all. They know the street corners to avoid because of drug dealing, and they know the schools that are in good repair and those with crumbling paint and broken windows.

High-resolution spatial and temporal data such as characteristics of householders, property owners, and neighborhood quality are important to the development of the NSDI and to other federal data initiatives. Local data are needed to complement federal data to address issues such as crime prevention, growth management, and land use decisions (NRC, 2002c). Demonstration projects conducted by the FGDC, the National Partnership for Reinventing Government, and five other federal agencies in 2000 found that federal standards can enhance a given community's ability to acquire comparable data from neighboring communities (FGDC, 2000). HUD can encourage and support the recipients of HUD funding so that their data conform to FGDC standards, are comparable across the nation, and can be used to update national databases.

Conclusion: HUD can use GIS as a tool to strengthen the agency's commitment to engaging communities by taking advantage of local knowledge about housing and urban development through more effective public participation in the development of local data sets. This is notably important for HUD's efforts in the colonias (Box 3.1) where local housing and development issues can be understood only in the context of labor and property markets operating at multiple levels from local to international.

Recommendation: HUD should facilitate the integration of local datasets and the development of mapping applications using the shared data; encourage public participation in the development and use of local data sets; and partner to develop local and in-house GIS capability.

Cooperative, Voluntary Relationships

Many of HUD's relationships are voluntary and cooperative. These include relationships with other federal agencies for data acquisition and dissemination. Voluntary relationships occur because both sides desire them and agree to them. If cooperative, voluntary relationships are to exist, each partner must find something of value in the relationship. A description of several of HUD's voluntary relationships follows.

The E-Government Act of 2002

The major goal of E-Gov[1] is to integrate management performance and budget through the development of an electronic government. Sometimes called Electronic or E-government, E-Gov was created by the E-Government Act of 2002 and has the potential to increase agency productivity, reduce redundancy, and improve service delivery. The act strives to bring together all levels of government, the private sector, and non-governmental organizations in an effort to integrate data across institutional boundaries in service to the common good. Realizing that, these goals require community-level participation to help build national data resources from the bottom up. This will include elements of statutory, hierarchical relationships as well as cooperative, voluntary relationships.

Sharing Information in Voluntary Relationships

On what basis do organizations share information when such activity is voluntary? First, organizations share information when each has something

---

[1]See <http://www.e-gov.com/>.

that the other organization desires, and the groups can use complementarities as the basis for a *quid pro quo*. A second reason is a sense of professional responsibility. In the case of federal agencies, all share a common responsibility to promote the welfare of the people of the United States. In particular, federal agencies with data collection mandates have the responsibility to contribute to the NSDI.

Efforts to share data can encounter barriers. Organizations may be at varying stages in their development of GIS, and in building their datasets for their own institutional needs. Some organizations have fully functional, comprehensive GIS; others are still working to achieve this. Still others are content to maintain the *status quo*, or are without the financial ability to make the leap to GIS. When organizations at different stages of GIS development come together, achieving cooperation may be difficult even if the parties are willing in principle to share data.

As HUD develops its GIS initiatives, it will have to understand the many and varied voluntary and cooperative arrangements that will help it succeed. Although, in many instances, the agency may be able to establish *quid pro quo* with other organizations, this will not always be possible. In other cases, especially in its work with small local and neighborhood organizations, HUD may encounter challenges that will require creative strategies to advance its GIS aspirations (Obermeyer, 1995). Addressing privacy considerations to encourage data sharing is one of these challenges.

HUD Client Groups (a Special Relationship)

HUD was established for the purpose of making housing affordable, revitalizing urban areas, and encouraging home ownership. HUD's mission is challenging largely because it embodies issues of scale. The overall force of HUD's institutional mission suggests that cities are HUD's primary client, yet, in fulfilling its mission, HUD directly affects smaller communities, individuals, and families.

Responsibility between HUD and its client groups flows both ways. HUD has responsibility toward the client groups, and these client groups in return can be a significant source of support for HUD. Each organization that comes into being and flourishes does so because it identifies and develops relationships with specific groups of people who benefit from the efforts of the organization (Obermeyer, 1990, 49). HUD provides its clients with a means to access "the system" in order to promote their objectives. GIS at the community level provides an opportunity to strengthen this link.

Data and GIS Intermediaries

Data and GIS intermediaries provide technical assistance or training in the use of geographic data. The work of these intermediaries demonstrates collaborations developed in the process of creating city and metropolitan data centers. Intermediaries provide a setting that enables groups to gather, clean, and present data in a comprehensive, timely, and detailed manner. Data and GIS intermediaries can bring public and private data together and build on local knowledge. Such partnerships can increase awareness of environmental justice issues such as the location of hazardous waste sites and transportation routes for the waste in relation to residential areas. An example is the Urban Strategies Council in Oakland, California,[2] which integrates 19 agencies to share data and create maps to reduce poverty and help transform low-income neighborhoods into healthy and thriving communities.

HUD's Relationship with States

HUD works primarily with groups and individuals at the local and metropolitan level, but states also have an interest in seeing that cities and communities within their borders maintain their viability. In this common interest shared by HUD, the 50 states, and localities across the nation are the seeds of a relationship that includes an expanded role for states. HUD provides funds to states through Community Development Block Grants (CDBG) and the Home Investment Partnerships Program (HOME), but also through the agency's Rural Housing and Economic Development program, Emergency Shelter Grants, and Housing Opportunities for Persons with Aids (HOPWA) funds. HOME funds create affordable housing for low-income families. The CDBG program, traditionally focused on affordable housing, now targets economic development activities (job and business development) in low-income neighborhoods. Similarly, the rural development program funds promote jobs, housing construction, and business development.

Community development efforts are facilitated by establishing and strengthening local, non-governmental agencies in metropolitan areas and by a strong role for federal and state governments as supporting partners (Kingsley et al., 1997). Because the statutory and hierarchical links between HUD and states remain relatively weak, the agency will have to be creative in enhancing its relationship with the states. One way for HUD to strengthen its

---

[2]See <http://www.urbanstrategies.org/> for details.

relationships with states is to partner with the Department of Transportation (DOT). DOT's relationships with state governments through the state DOTs is strong, and its mission—to promote access to transportation to enhance the quality of life in the United States—intersects with HUD's mission to support community development and increase access to affordable housing.[3]

HUD may also find it valuable to strengthen relationships with state-level actors such as the state I-Teams,[4] for example, in New Jersey where the I-Team is building a statewide digital parcel-level layer. This effort includes state health data and incorporates neighborhood efforts to conduct parcel surveys recording local conditions (Hank Garie, New Jersey I-Team coordinator, personal communication, 2002). HUD will have to make a deliberate voluntary effort to expand its influence at the state level if it wishes to work more extensively with state agencies in the future.

Partnerships with Universities

HUD can use GIS as a tool to strengthen the agency's understanding and analysis of housing and urban issues through the agency's support of the work of university- and non-university-based research partners. Examples of ongoing work include HUD's international efforts with the University Consortium for Geographic Information Sciences (UCGIS) in urban indicators (Box 5.2) and community development grants to historically black colleges and universities (HBCU). The HBCU grants are used to revitalize communities surrounding the universities through estate acquisition, demolition, and rehabilitation; to provide homeownership assistance to low- and moderate-income persons; and for community economic development.

HUD also provides grants to university researchers, for example, to low-income graduate students in community development, urban planning, public policy, and public administration. Working with academic researchers, HUD can incorporate new methods of spatial analysis into the agencies research agenda and pass on the benefits of new tools and methods to partners and clients. HUD can use GIS to integrate the methodological expertise housed in universities and the local expertise of communities and local governments. At the same time, with HUD as a conduit to local community groups, academic researchers can gain access to local data and local knowledge about relevant urban and housing issues. Localities and local governments can develop research questions, analyze research results, and use the results of multivariate techniques to develop appropriate local solutions.

---

[3]<http://www.dot.gov/mission.htm>.
[4]Implementation-Teams.

BOX 5.2
Global Urban Indicators

Global Urban Indicators is a joint project between HUD and the University Consortium of Geographic Information Sciences (UCGIS) to "establish and develop an international collaborative network between American universities and universities in the developing world using GIS-based urban indicators for urban policy analysis" (UCGIS RFP 0002). One of the major goals of the initiative is to establish mechanisms for the systematic collection of data for more than 100 indicators for major cities in the developing world in format comparable to PD&R's State of the Cities data. The program links to the United Nations Habitat Program through its urban indicators project.

The program issued a draft report to HUD in May 2002 that identified "difficulty in reconciling the need for uniform and consistent urban indicators from the top down or global perspective, while developing a bottom-up perspective of developing useful data for local planning and policy analysis" (Dueker and Jampoler, 2002). The authors recommend building local capacity to collect and use policy-related indicators, thus increasing capacity of participants to continue urban indicator analysis on a long-term basis. Proposed next steps include comparison of the experience and results of analyzing urban indicators using GIS, assessment of strengths and weaknesses of various measures and approaches, and exploration of opportunities for increasing consistency.

Conclusion: HUD, through PD&R, can be the liaison between local governments and community groups and academic researchers to further the agency's research agenda and to define new ways for researchers to extend their skills to building local capacity and addressing local needs.

Recommendation: PD&R should build relationships with university and unaffiliated researchers to engender participation of local groups in policy analysis, research, and community building; and to promote the use of advanced spatial analysis in urban housing policy research to address the complexities of modern urban dynamics.

HUD's Federal Partners

HUD is an important link between the federal government and communities. The agency has an important role to play in ensuring that local data and community priorities are included in federal data initiatives and those concerning housing and urban development. Urbanization is on the rise at an unprecedented rate (Brennan, 1999). In 20 years, more than 60 percent of the world's population will be living in cities and that rate is even higher in the United States (Bugilarello, 2001). Cities and particularly neighborhoods are vitally important organizing units of human behavior, including economic activity and environmental impacts. Urban and neighborhood-level data will be needed to address issues ranging from water and air quality to crime and security. No other federal agency has HUD's history of contact and connection with cities and communities across the nation.

HUD's primary federal data-sharing partner is the U.S. Bureau of the Census, but there are many others including the federal Environmental Protection Agency, Health and Human Services, the Department of Justice, and the Federal Emergency Management Agency (see Table 5.1). Cooperating with other federal efforts will bring HUD as an institution into ongoing dialogue with other federal agencies and important state and local partners. This will ensure that HUD knows and follows federal data procedures and fully employs the developing spatial data infrastructure to benefit HUD as an agency, HUD's clients, and communities across the country.

TABLE 5.1   HUD's Federal Partners and Potential Areas of Collaboration

Agencies with Urban/Community Mandates	Potential Data Collaboration Areas
EPA	Environmental health, growth management, environmental justice
DOT	Public transit access, transportation planning, community development, welfare to work
FEMA	Emergency evacuation routes, Homeland Security
HHS	TANF information, addiction treatment and mental health services locations, Head Start services, Meals on Wheels
BLS	Unemployment and layoff rates, labor force status of persons by residence, jobs and wages by place of work, price and living conditions

Conclusion: GIS provides HUD with an opportunity to insert housing and urban issues onto the national agenda. HUD can partner with other federal agencies with responsibility for providing and managing data relevant to urban, community, and housing issues. Transportation, social services, and employment are intimately linked to housing and community development.

Recommendation: PD&R should take the lead in HUD in building interagency relationships with federal data-providing agencies that have responsibilities related to urban and community issues, notably the Department of Transportation, the Department of Health and Human Services, and the Environmental Protection Agency.

SUMMARY

Partnerships can facilitate many of HUD's goals described in this report. Developing, maintaining, and disseminating reliable spatial data is a major challenge for HUD.   The agency's efforts will be more efficient and productive when carried out in coordination with other federal data initiatives, notably the NSDI. The creation of an urban spatial data infrastructure as a component to the NSDI is a major recommendation of this committee. Achieving this goal will depend on relationships among the federal agencies with data collection and dissemination responsibilities, and with local groups that are the best sources of relevant, accurate local data.

The needs of HUD's data users and stakeholders in urban and community development are best identified in the context of ongoing relationships for data collection and best addressed by building local capacity for spatial analysis. Support is essential for local governments and other local-level users to develop capability in spatial analytical research and for more advanced research taking place inside HUD and in universities and other urban research centers. GIS tools, such as an online clearinghouse for spatial data research and urban simulation models, can be used to promote analysis of complex urban issues spanning geographic scales of neighborhood, community, region, state, and nation, which are at the heart of HUD's mission.

In summary, HUD can use GIS to work through interagency groups to achieve consensus on housing and urban issues and promote a coherent urban development policy in the United States in the following ways:

1. HUD can use GIS to facilitate the agency's efforts to interact with organizations beyond its institutional boundaries. GIS can facilitate the development of vertical and horizontal networks to share data,

discuss housing and urban issues, and ultimately create public policy to respond to these issues.

2. HUD can use GIS as a tool to strengthen HUD's commitment to engaging communities by taking advantage of local knowledge about housing and urban development through more effective public participation.

3. HUD can use GIS as a tool to strengthen the agency's understanding and analysis of housing and urban issues by supporting the work of university and non-university-based researchers to develop multivariate spatial analysis techniques.

4. HUD can use GIS to integrate the methodological expertise housed within universities with the local expertise of communities and local governments. Localities and local governments can develop research questions, analyze research results, and use research findings to develop appropriate local solutions.

5. GIS can help HUD to think about housing and urban issues at different scales. Building relationships with local data producers and actors can facilitate this process.

6. GIS offers HUD an opportunity to insert housing and urban issues onto the national agenda by participating fully in the federal data initiatives including the FGDC, NSDI, and E-Gov.

HUD can further these goals by fostering partnerships among local groups and urban researchers, and by cultivating relationships with other federal agencies that share HUD's responsibility for the future of American cities and the well-being of those who live there.

References

Aronoff, S. 1989. Geographic Information Systems: A Management Perspective. Ottawa, ON, Canada: WDL Publications.

Barndt, M. 2002. Initiative on GIS and Society. Available at: <http://www.geo.wvu.edu/i19/papers/barndt.html>.

Brennan, E. 1999. Population, Urbanization, Environment, and Security: A Summary of Issues. Washington, D.C.: Woodrow Wilson International Center for Scholars.

Brockerhoff, M. 2000. An Urbanizing World. Population Bulletin 55 (3). Available at: <http://www.prb.org/Content/NavigationMenu/PRB/AboutPRB/Population_Bulletin2/An_Urbanizing_World.htm>.

Brophy, P. C., and R. N. Smith. 1997. Mixed Income Housing: Factors for Success. CityScape: A Journal of Policy Development and Research 3(2):3-31.

Bugilarello, G. 2001. Rethinking Urbanization. The Bridge 32(1):5-12.

Castells, M. 1996. The Rise of the Network Society. Vol. 1 of The Information Age: Economy, Society, and Culture. Oxford, England, and Malden, MA: Blackwell Press.

Cisneros, H. C. 1996. Regionalism: The New Geography of Opportunity. CityScape: A Journal of Policy Development and Research, Special Issue, December:35-53.

Citro, C. F., and R. T. Michael, eds. 1995. Measuring Poverty: A New Approach. Washington, D.C.: National Academy Press.

Crane, J. 1991. The Epidemic Theory of Ghettos and Neighborhood Effects of Dropping out of High School and Teen Age Childbearing. American Journal of Sociology 96(5):1126-1159.

de la Barra, T. 1995. Integrated Land Use and Transportation Modeling: Decision Chains and Hierarchies. Cambridge: Cambridge University Press.

de Leeuw, F., and R. J. Struyk. 1975. The Web of Urban Housing: Analyzing Policy with a Market Simulation Model. Washington, D.C.: Urban Institute.

Downs, A. 1981. Neighborhoods and Urban Development. Washington, D.C.: Brookings Institution.

Dueker, K., and S. Jampoler. 2002. Final Report to U.S. Department of Housing and Urban Development Global Urban Quality: An Analysis of Urban Indicators Using Geographic Information Science, May 2002, UCGIS, draft. Available at: <http://www.cobblestoneconcepts.com/ucgis2hud/FinalHudReport.htm# Conclusions and Recommendations>.

Echenique, M. 1994. Urban and Regional Studies at the Martin Centre: Its Origins, Its Present and Its Future. Environment and Planning B 21:517-533.

Ellen, I. G. 1999. Spatial Stratification within U.S. Metropolitan Areas. Pp. 192-212 in Altshuler, A., W. Morrill, H. Wolman and F. Mitchell, eds., Governance and Opportunity in Metropolitan America. Washington, D.C.: National Academy Press.

Ellen, I. G., M. Schill, S. Susin, and A. Schwartz. 2001. Building Homes, Reviving Neighborhoods: Spillovers from Subsidized Construction of Owner-Occupied Housing in New York City. Journal of Housing Research 12(2):185-216.

Farley, R., and W. Frey. 1994. Changes in the Segregation of Whites from Blacks during the 1980s: Small Steps towards a More Integrated Society. American Sociological Review 59(February):23-45.

FGDC [Federal Geographic Data Committee]. 2000. NSDI Community Demonstration Projects Final Report. Available at: <http://www.fgdc.gov/nsdi/docs/cdp/>.

GAO [Government Accounting Office]. 2001. HUD Information Systems: Immature Software Acquisition Capability Increases Project Risks (GAO-01-962). Washington, D.C.: Government Printing Office.

Goering, J., H. Elhassan, J. Feins, M. J. Holin, J. Kraft, and D. McGinnis. 1999. Moving to Opportunity for Fair Housing Demonstration Program: Current Status and Initial Findings. Washington, D.C.: U.S. Department of Housing and Urban Development.

Hanushek, E., and C. F. Citro, eds. 1991. Improving Information for Social Policy Decisions: The Uses of Microsimulation Modeling, vol. 1 and 2. Washington, D.C.: National Academy Press.

Hasson, J. 2000. Public Gets Access to HUD Database. Federal Computer Week, Nov. 22. Available at: <http://www.fcw.com/civic/articles/2000/1120/web-hud-11-22-00.asp>.

HUD [U.S. Department of Housing and Urban Development]. 1998. Mapping Your Community: Using Geographic Information to Strengthen Community Initiatives. Washington, D.C.: HUD.

HUD [U.S. Department of Housing and Urban Development]. 1999. Building Excellence in Spatial Analysis: PD&R's Management Strategy in Support of Geographic Information Analysis. Washington, D.C.: HUD.

HUD [U.S. Department of Housing and Urban Development]. 2000. Strategic Plan FY2000-FY 2006. Washington, D.C.: HUD.

HUD [U.S. Department of Housing and Urban Development]. 2002a. Mission and History. Washington, D.C.: HUD. Available at <http://www.hud.gov/library/bookshelf18/mission.cfm>.

HUD [U.S. Department of Housing and Urban Development]. 2002b. Draft Strategic Plan, 2003-2008. Washington, D.C.: HUD.

Jankowski, P., and T. Nyerges. 2001. Geographic Information Systems for Group Decision Making: Toward a Participatory, Geographic Information Science. London: Taylor and Francis.

Jargowsky, P. 1997. Poverty and Place: Ghettos, Barrios, and the American City. New York: Russell Sage Foundation.

Jencks, C., and S. E. Mayer. 1990. Social Consequences of Growing Up in a Poor Neighborhood. Pp. 111-186 in L. E. Lynn, Jr., and M. G. H. McGeary, eds., Inner-City Poverty in the U.S. Washington, D.C.: National Academy Press.

Johnson, M. P. 2002. Decision Support for Family Relocation Decisions under the Section 8 Housing Assistance Program Using Geographic Information Systems and Analytic Hierarchy Process. Journal of Housing Research 12(2):277-306.

Kain, J. F., and W. C. Apgar, Jr. 1985. Housing and Neighborhood Dynamics: A Simulation Study. Cambridge, MA: Harvard University Press.

Khadduri, J., and M. Martin. 1997. Mixed-Income Housing in the HUD Multi-Family Stock. CityScape: A Journal of Policy Development and Research 2:33-69.

Kingsley, T. G., J. B. Mc Neely, and J. O. Gibson. 1997. Community Building Coming of Age. Washington D.C.: Urban Institute.

Kingsley, G. T., J. Johnson, and K. L. S. Pettit. 2001. HOPE VI and Section 8: Spatial Patterns in Relocation. Washington, D.C.: Urban Institute.

Lichter, D., and M. Crowley. 2002. Poverty in America: Beyond Welfare Reform. Population Bulletin 57(2), June issue.

Longley, P. A., M. F. Goodchild, D. J. Maguire, and D. W. Rhind. 2001. Geographic Information Systems and Science. New York: John Wiley and Sons.

Massey, D. S. 1990. American Apartheid: Segregation and the Making of an Underclass. American Journal of Sociology 96:329-357.

McDowell, B. D. 2001. Evaluating Methods for Monitoring and Improving HUD-Assisted Housing. Washington D.C.: National Academy of Public Administration.

Nelson, K. P. 2001. What Do We Know about Shortages of Affordable Housing? Testimony before the House Committee on Financial Services, Subcommittee on Housing and Community Opportunity Office of Policy Development and Research, U.S. Department of Housing and Urban Development. Available at <http://financialservices.house.gov/media/pdf/ 050301nepdf>.

NRC [National Research Council]. 1993. Toward a Coordinated Spatial Data Infrastructure for the Nation. Washington, D.C.: National Academy Press.

NRC [National Research Council]. 1997. The Future of Spatial Data and Society—Summary of a Workshop. Washington, D.C.: National Academy Press.

NRC [National Research Council]. 1998. People and Pixels: Linking Remote Sensing and Social Science. Washington, D.C.: National Academy Press.

NRC [National Research Council]. 1999. Distributed Geolibraries—Spatial Information Resources. Washington, D.C.: National Academy Press.

NRC [National Research Council]. 2002a. IT Roadmap to a Geospatial Future. Washington, D.C.: National Academy Press.

NRC [National Research Council]. 2002b. Research Opportunities in Geography at the U.S. Geological Survey. Washington, D.C.: National Academy Press.

NRC [National Research Council]. 2002c. Community and Quality of Life: Data Needs for Informed Decision Making. Washington, D.C.: National Academy Press.

NRC [National Research Council]. 2002d. Measuring Housing Discrimination in a National Study. Washington, D.C.: National Academy Press.

NSGIC and FGDC [National States Geographic Information Council and Federal Geographic Data Committee]. nd. Using Geography to Advance the Business of Government: The Power of Place to Support Decision Making. CD-ROM. Washington, D.C.: NSGIC.

Obermeyer, N. J. 1990. Bureaucrats, Clients and Geography. Geography Research Paper No. 216. Chicago: University of Chicago Press.

Obermeyer, N. J. 1995. Reducing Interorganizational Conflict to Facilitate Sharing Geographic Information. Pp. 138-148 in Sharing Geographic Information, H. J. Onsrud and G. Rushton, eds. New Brunswick, NJ: Center for Urban Policy Research.

Obermeyer, N. J., and J. Pinto. 1994. Managing Geographic Information Systems. New York: Guilford Press.

Office of Technology Assessment. 1995. The Technological Reshaping of Metropolitan America. Washington, D.C.: Government Printing Office.

Orcutt, G. 1957. A New Type of Socio-Economic System. Review of Economics and Statistics 58:773-797.

Orcutt, G. 1960. Simulation of Economic Systems. American Economic Review, December:893-907.

Orfield, M. 1997. Metropolitics: A Regional Agenda for Community and Stability. Washington, D.C.: Brookings Institution.

Park, R. E. 1926. The Urban Community as a Spatial Pattern and a Moral Order. Pp. 21-31 in C. Peach, ed., Urban Social Segregation. London, UK: Longman Press.

Pickles, J. 1995. Ground Truth: The Social Implications of Geographic Information Systems. New York: Guilford Press.

Putnam, S. 1983. Integrated Urban Models. London: Pion Publishers.

Rubinowitz, L. S., and J. E. Rosenbaum. 2000. Crossing the Class and Color Lines: From Public Housing to White Suburbia. Chicago: University of Chicago Press.

Rusk, D. 1999. The Inside/Outside Game: Winning Strategies for Saving Urban America. Washington, D.C.: Century Foundation.

Schelling, T. C. 1971. Dynamic Models of Segregation. Journal of Mathematical Sociology 1:143-186.

Schelling, T. C. 1972. The Process of Residential Segregation: Neighborhood Tipping. Pp. 157-184 in Racial Discrimination in Economic Life, A. Pascal, ed. Lexington, MA: Lexington Books.

Schelling, T. C. 1978. Micromotives and Macrobehavior. New York: W. W. Norton.

Schill, M. H., and G. Daniels. 2002. The State of New York City Housing and Neighborhoods: an Overview of Recent Trends. Paper and Policies to Promote Affordable Housing Conference, February 27, 2002. New York: Federal Reserve Bank.

Sen, A. 1976. Famines as Failures of Exchange Entitlements. Economic and Political Weekly, Special Issue, August.

Smith, N. 1996. The New Urban Frontier: Gentrification and the Revanchist City. New York and London: Routledge.

Thompson, D., and W. Sherwood. 1999. Spatial Analysis of the Location of Households Living in HUD-Assisted Low-income Housing. Washington, D.C.: U.S. Department of Housing and Urban Development.

Turner, M. A., S. Popkin, and M. Cunningham. 1999. Section 8 Mobility and Neighborhood Health: Emerging Issues and Policy Challenges. Washington, D.C.: Urban Institute.

Waddell, P. (2002). UrbanSim: Modeling Urban Development for Land Use, Transportation and Environmental Planning. Journal of the American Planning Association 68(3):297-314.

Waddell, P. 1994. Housing Conditions in the City of Dallas: Comprehensive Technical Report. Dallas: University of Texas.

Waddell, P., B. Berry, and I. Hoch. 1993. Residential Property Values in a Multinodal Urban Area: New Evidence on the Implicit Prices of Location. Journal of Real Estate Finance and Economics 7(2):117-143.

Wilson, W. J. 1987. The Truly Disadvantaged: The Inner City, the Underclass, and Public Policy. Chicago: University of Chicago Press.

Wu, X. B., and D. Z. Sui. 2001. An Initial Exploration of a Lacunarity-based Segregation Measure. Environmental and Planning B 28(3):443-446.

Wyly, E. K., and S. R. Holloway. 2002. The Disappearance of Race in Mortgage Lending. Economic Geography 78(2):129-163.

Appendixes

Appendix A

Biographical Sketches of Committee Members and Staff

Eric Anderson, *Chair,* is the Chief Executive Officer (CEO) and city manager of Des Moines, Iowa. As the CEO for the city, he has been active in using GIS capabilities to integrate and coordinate all municipal functions. As city manager, he has developed an enterprise in GIS platform to integrate citizen input and other city data in a geo-linked system. Mr. Anderson received an M.A. from the Maxwell School at Syracuse University and a Master of Public Administration degree from the Graduate School of Public Affairs, University of New York at Albany. He is a fellow of the National Academy of Public Administration (NAPA) and serves as assistant director for research and development of the International City and County Managers Association. Mr. Anderson was a member of the 1998 NAPA study on Geographic Information for the 21st Century and has been active in coordination activities for the national spatial data infrastructure. In addition, he is an active member of the NRC's Mapping Science Committee.

Nina S.-N. Lam is Richard J. Russell Louisiana Studies Professor in Geography, Department of Geography and Anthropology, Louisiana State University. Her research interests include cartography, GIS, remote sensing, quantitative methods, medical geography, and China. She has served on the NRC's Mapping Science Committee. She is currently involved in studies that use census, environmental data, and public health information to examine spatial relationships between disease clustering and environmental factors. She has compiled indexes of regional economic development characteristics and used demographic data to determine voting patterns.

Kathe A. Newman is a Lecturer at Bloustein School of Planning and Public Policy, Center for Urban Policy Research, Rutgers University. Dr. Newman's research interests include urban politics, qualitative research methods, and community development and the intersection of race, ethnicity, gender, and class. She has been working with neighborhood-level data and indicators and has used microlevel datasets to empower neighborhoods. She is a proponent for democratization of data in action.

Tim Nyerges is a Professor in the Department of Geography at the University of Washington in Seattle. His research includes GIS, spatial decision support systems and group decision making, transportation and environmental analysis using GIS, human-computer interaction, and spatial cognition. He earned his Ph.D. in GIS from Ohio State University. Dr. Nyerges has developed a framework for gathering information for improved decision making based on diverse data-gathering strategies to result in more creative uses of GIS. He has experience in building GIS software and in the policy implications of applying GIS.

Nancy J. Obermeyer is Associate Professor of Geography at Indiana State University. Her research interests have emphasized the institutional and societal ramifications of implementing GIS. Specifically, her focus has addressed two separate conceptual issues: how the proliferation of GIS databases and differential access to spatial databases influences the abilities of different social groups to use information for their own empowerment; and the possibilities and limitations of using GIS as a participatory conflict resolution tool. She has also been involved with the development of certification for GIS professionals, and is co-author (with Jeffrey Pinto) of Managing GIS (Guilford 1994, 2004). In an earlier career in public service, Dr. Obermeyer held professional assignments with the State of Illinois in the areas of planning, transportation, and energy.

Myron Orfield is Executive Director of the Metropolitan Area Research Corporation, Minneapolis, Minnesota. His specific areas of interest include urban sprawl, regional governance, transit, taxes, and environment. He is the president and founder of the Metropolitan Area Research Corporation (MARC), a non-profit research and GIS firm. MARC's objective is to study the growing social and economic disparity and inefficient growth patterns in metropolitan areas and assist individuals and groups in promoting greater equity, local reinvestment, environmental protection through land-use planning, responsive regional-level decision making, and reduction in poverty. He was a participant in the American Planning Association's Growing Smart Project. Mr. Orfield served on the Committee on Improving the Future of U.S. Cities through Improved Metropolitan Area Governance. He has served five terms in the Minnesota House of Representatives and is currently the State Senator from southwest Minneapolis. Mr. Orfield did graduate work at Princeton University and has a law degree from the University of Chicago. He has practiced in the private sector and currently teaches at the University of Minnesota Law School.

John Pickles is a Professor of Geography and the Earl N. Phillips Distinguished Chair in International Studies at the University of North Carolina at Chapel Hill. His research focus includes the way geographers understand science, space, and everyday life; political and cultural economy; geographic thought and social theory philosophy of science; political economy; regional development; and transition theory and democratization. He has studied the role of technological science in constructing rationalized and mapped worlds and has written about the social implications and limitations of GIS.

Daniel Z. Sui is a Professor of Geography and holds the Reta A. Haynes endowed chair at Texas A&M University, College Station. His primary research interests include the integration of spatial analysis and modeling with GIS for socioeconomic and environmental applications, theoretical issues in geographic information sciences, and the emerging geographies of the information society. Dr. Sui teaches GIS courses, GIS-based spatial analysis and modeling, and advanced research seminars in urban and economic geography. He aims to make GIS a more robust decision-making tool, and has studied the integration of GIS and urban modeling, and GIS with remote sensing data to monitor land-use changes. Dr. Sui earned his Ph.D. in Geography from the University of Georgia.

Paul A. Waddell is Associate Professor of Public Affairs and Associate Professor of Urban Design and Planning at the University of Washington. His research interests include transportation planning, urban development, GIS, and infrastructure planning. He has recently been guiding the development of a new urban simulation model for metropolitan policy and planning down to the parcel-level scale, to be used to make residential location choices, transportation and real estate development decisions, and price setting. He has a strong interest in connecting the model to housing policy questions. He earned his Ph.D. in political economy from the University of Texas at Dallas and his dissertation topic related to factors determining household choice.

National Research Council Staff

Lisa M. Vandemark has a Ph.D. in Geography from Rutgers University and a M.S. in Human Ecology from the University of Brussels, Belgium. Her B.S. (nursing) is also from Rutgers University. Currently she is a Program Officer at the NRC's Board on Earth Sciences and Resources. Prior to this appointment she was a research associate at the Institute of Marine and Coastal Sciences, Rutgers University, and an intern at the National Science Resources Center at the Smithsonian Institution. Her research interests include environmental perception and decision making, international development, natural resource management, and the role of interdisciplinary studies in environmental protection.

Monica R. Lipscomb is a Research Assistant for the NRC's Board on Earth Sciences and Resources. She has completed coursework for a Master's degree in Urban and Regional Planning at Virginia Polytechnic Institute, with a concentration in Environmental Planning and graduate certificate in International Development. Previously she served as a Peace Corps volunteer in Côte d'Ivoire and has worked as a biologist at the National Cancer Institute. She holds a B.S. in Environmental and Forest Biology from the State University of New York–Syracuse.

Appendix B

Workshop Agenda and Participants

AGENDA

Workshop for Committee to Review Research and Applications of GIS at the U.S. Department of Housing and Urban Development (HUD)
National Academy of Sciences Building
2101 Constitution Avenue, N.W., Members Room
Washington, D.C.
April 25-27, 2002

Thursday, April 25

OPEN SESSION

6:30 p.m.   Welcome Dinner for Participants and Guests

7:00 p.m.   GI for America
             Ron Matzner, Federal Geographic Data Committee

Cecil and Ida Green Building
2001 Wisconsin Avenue, N.W., Room 110
Washington D.C.

Friday, April 26

CLOSED SESSION

8:00 a.m.   Brief discussion of meeting plans

8:45   Welcome
                 Eric Anderson, Chair

9:00   Keynote Address
                 I-Team Progress in New Jersey
                 Hank Garie, New Jersey GIS I-team Coordinator

10:00 -10:15   Break

10:15-12:00   Panel on Neighborhood Change

                 Spatial dimensions of poverty
                 Paul Jargowsky, University of Texas at Dallas

                 Geographies of Mortgage Market Segregation In Essex
                     County, New Jersey
                 Elvin Wyly, Rutgers University

                 GIS, Neighborhood Change and Racial Dynamics
                 Alexander Von Hoffman, Harvard University

                 Community building with GIS
                 Josh Kirschenbaum, PolicyLink

12:00 p.m.   Lunch

1:00-3:00 p.m.   Low-Income Housing Needs/Availability Analysis

                 Location, Location, Mobility
                 Steve Redburn, Housing Division of OMB

Applications of GIS to Fair Housing
Stella Adams, North Carolina Affordable Housing Center

Low-income housing data
Laura Harris, Urban Institute

Experiences From the Field: How GIS is Being Used, or
    Not Being Used, in HUD Field Offices
Michael Martin, U.S. HUD, Milwaukee, Wisconsin

Affordable Housing
Kathryn Nelson, U.S. HUD, Washington, D.C.

3:00-3:15   Break

3:15-5:00   Panel on the Colonias: Border Issues and GISciences

GIS needs in the Colonias for Occupancy Inventories
Jeremiah Carew, PriceWaterhouseCoopers

Community Based GIS Development in U.S./Mexico
    Border Communities: The Colonias Monitoring Project
Sherry Durst, USGS

Colonias Inventory for HUD
Bob Czerniak, New Mexico State University

Spatial and Demographic Data for Colonias
Michael Ratcliffe, U.S. Census Bureau

5:00 p.m.   Adjourn

Saturday, April 27

8:00-9:45 a.m.   Metropolitan and Regional-Level, Multi-scale
Analysis

Public Housing and Geographies of Violent Crime: Issues
    of Scale and Context
Steven Holloway, University of Georgia

National Land Monitoring System
Gerrit Knaap, University of Maryland

Land Use Trends and Segregation
David Rusk, Urban Policy Consultant

Growth Management and GIS
Emily Talen, University of Illinois at Urbana-Champaign

9:45- 10: 00   Break

10:00 –11:30   Data Applications

GIS Data Applications and Interoperability
Myra Bambacus, NASA/FGDC

Beyond Crime Mapping
Luc Anselin, University of Illinois at Urbana-Champaign

Environmental Justice: Spatial Analysis of Health and
        Housing
Juliana Maantay, City University of New York

Supporting Programs and Policies Relevant to Urban
        Neighborhoods: The Role of Data & GIS Services
Michael Barndt, Nonprofit Center of Milwaukee

11:30   Round-Table Discussion: Common Threads and
                        Emerging Themes
                Committee, workshop participants, and guests.

12:00 p.m.   Lunch

1:00-5:00 p.m.   **CLOSED SESSION**

Appendix C

List of Contributors

Stella Adams, North Carolina Affordable Housing Center, Durham

Luc Anselin University of Illinois at Urbana-Champaign, Urbana, Illinois

Myra Bambacus, National Aeronautics and Space Administration, Goddard Space Flight Center, Greenbelt, Maryland

Michael Barndt, Nonprofit Center of Milwaukee, Wisconsin

Dick Burke, Deputy CIO for IT Reform, U.S. Department of Housing and Urban Development, Washington, D.C.

Jeremiah Carew, PriceWaterhouseCoopers, Arlington, Virginia

David Chase, Policy Development & Research, U.S. Department of Housing and Urban Development, Washington, D.C.

Bob Czerniak, New Mexico State University, Las Cruces

Sherry Durst, U.S. Geological Survey, Denver, Colorado

Hank Garie, New Jersey GIS I-team Coordinator, Trenton

Laura Harris, Urban Institute, Washington, D.C.

Steven Holloway, University of Georgia, Athens, Gorgia

Paul Jargowsky, University of Texas at Dallas, Richardson, Texas

Josh Kirschenbaum, PolicyLink, Oakland, California

Gerrit Knaap, University of Maryland, College Park, Maryland

Alven Lam, Project Manager, U.S. Department of Housing and Urban Development's Office of International Affairs, Washington, D.C.

Juliana Maantay, City University of New York, New York

Michael Martin, U.S. Department of Housing and Urban Development, Milwaukee, Wisconsin

Ron Matzner, Federal Geographic Data Committee, Reston, Virginia

Mark Mitchell, Project Manager, Environmental Systems Research Institute, Redlands, California

Kathryn Nelson, U.S. Department of Housing and Urban Development, Washington, D.C.

Michael Ratcliffe, U.S. Census Bureau, Washington, D.C.

Steve Redburn, Housing Division of Office of Management and Budget, Washington, D.C.

Todd Rogers, Federal Marketing Manager, Environmental Systems Research Institute, Redlands, California

David Rusk, Urban Policy Consultant, Washington, D.C.

Ayse Can Talen, Senior Director, Community Development Research, Fannie Mae Foundation, Washington, D.C.

Emily Talen, University of Illinois at Urbana-Champaign, Champaign, Illinois

Alexander Von Hoffman, Harvard University, Cambridge, Massachusetts

Elvin Wyly, Rutgers University, Piscataway, New Jersey

Appendix D

List of Acronyms

ArcIMS	Arc Internet Map Server
AHS	American Housing Survey
BLM	Bureau of Land Management
CDBG	Community Development Block Grant
COPC	Community Outreach Partnerships Centers
CORSIM	Cornell Simulation Model
DIME	Dual Independent Map Encoding
DOT	U.S. Department of Transportation
DRAM	Disaggregated Residential Allocation Model
DYNASIM	Dynamic Simulation of Income Model
E-gov	Electronic government
EGIS	Enterprise Geographic Information Systems
E-MAPS	Environmental Maps
EMPAL	Employment Allocation Model
EPA	U.S. Environmental Protection Agency
ESRI	Environmental Systems Research Institute
FBI	Federal Bureau of Investigation
FEMA	Federal Emergency Management Agency
FGDC	Federal Geographic Data Committee
FHA	Federal Housing Assistance
GAO	General Accounting Office
GCDB	Geographic Coordinate Data Base
GIS	Geographic Information Systems

GML	Geographic Markup Language
GSDI	Global Spatial Data Infrastructure
GPS	Global Positioning System
HBCU	Historically black colleges and universities
HMDA	Home Mortgage Disclosure Act
HOME	Home Investment Partnerships Program
HOPWA	Housing Opportunities for Persons with Aids
HRA	Housing Relocation Assistant
HUD	U.S. Department of Housing and Urban Development
HUDS	Harvard Urban Development Simulation
ICT	Information and Communication Technology
ISTEA	Intermodal Surface Transportation and Efficiency Act
I-Team	Implementation team
LIHTC	Low-Income Housing Tax Credit
MATH	Micro-Analysis of Transfers to Households
MSAs	Metropolitan Statistical Areas
NBER	National Bureau of Economic Research
NKLA	Neighborhood Knowledge Los Angeles
NRC	National Research Council
NSDI	National Spatial Data Infrastructure
PD&R	Office of Policy Development and Research
POMS	Property Owners and Managers Survey
R-MAPS	Research maps
SDI	Spatial Data Infrastructure
SOCDS	State of the Cities Data systems
TAZ	Traffic Analysis Zone
TEA21	Transportation Equity Act
TIGER	Topologically Integrated Geographic Encoding and Referencing System
UCGIS	University Consortium for Geographic Information Science
USDI	Urban Spatial Data Infrastructure
USGS	U.S. Geological Survey